Diagnostic Molecular Pathology

A Practical Approach

Volume II
Cell and Tissue Genotyping

Edited by

C. S. HERRINGTON

and

J. O'D. McGEE

Nuffield Department of Pathology and Bacteriology
John Radcliffe Hospital
Oxford, OX3 9DU
UK

D0706292

————at————
OXFORD UNIVERSITY PRESS
Oxford New York Tokyo

Oxford University Press, Walton Street, Oxford OX2 6DP
Oxford   New York   Toronto
Delhi   Bombay   Calcutta   Madras   Karachi
Petaling Jaya   Singapore   Hong Kong   Tokyo
Nairobi   Dar es Salaam   Cape Town
Melbourne   Auckland
and associated companies in
Berlin   Ibadan

Oxford is a trade mark of Oxford University Press

A Practical Approach 🔬 is a registered trade mark
of the Chancellor, Masters, and Scholars of the University of Oxford
trading as Oxford University Press

Published in the United States
by Oxford University Press, New York

A catalogue record for this book is available from the British Library

Library of Congress Cataloging in Publication Data
Diagnostic molecular pathology v. 2 : a practical approach / edited by C. S.
Herrington and J. O'D. McGee.
(Practical approach series)
Includes bibliographical references and index.
1. Molecular probes—Diagnostic use.   2. Pathology, Molecular.
3. Molecular biology—Methodology.   I. Herrington, C. S.
II. McGee, J. O'D. (James O'Donnell)   III. Series.
[DNLM: 1. Genetics, Biochemical.   2. Molecular Biology.
3. Molecular Probe Techniques.   4. Pathology—methods. QZ 25 D5356]
RB43.7.D53   1992   616.07'56—dc20   91–35355
ISBN 0–19–963241–3 (set : spiral hardback)
ISBN 0–19–963237–5 (v. 1 : spiral hardback)
ISBN 0–19–963239–1 (v. 2 : H/B)
ISBN 0–19–963240–5 (set : pbk.)
ISBN 0–19–963236–7 (v. 1 : pbk.)
ISBN 0–19–963238–3 (v. 2 : P/B)

Set by Footnote Graphics, Warminster, Wilts
Printed in Great Britain by Information Press Ltd, Eynsham, Oxford

Preface

Over the past 20 years, there have been major advances in molecular pathology due to the introduction of monoclonal antibodies and nucleic acid technology. These techniques have revolutionized the way we practise laboratory medicine. Monoclonal antibodies have only been routinely used in most diagnostic laboratories over the past 10 years although they were discovered in 1975. It is not presumptious, therefore, to suppose that within a few years, nucleic acid technology will be in routine use in diagnostic laboratories particularly as non-radioactive methods for nucleic acid detection are simplified and become more sensitive. Both genotyping and phenotyping of intact and extracted cells have unravelled the mechanisms and causes of some disorders; for example, the molecular basis of genetic disorders and the diagnosis of HIV, HPV, and other infections. These technologies have also made inroads into the understanding of tumour biology and its prognosis. Many other developments are on the horizon which will lead to further improvements in laboratory diagnosis of disease and therefore clinical understanding and management.

The purpose of this book is to present a comprehensive view of most techniques used in diagnostic molecular pathology, their applications, and limitations. Each chapter is written by an expert in the field who has 'hands on' experience of the technology. The detailed protocols will help the novice and experienced investigator in approaching the clinical laboratory diagnosis of disease. In Volume I, we have covered two aspects of molecular pathology: phenotyping and genotyping of intact cells. There are two chapters on phenotyping by immunocytochemistry at the light and electron microscopic levels. The clinical applications of immunophenotyping in cancer diagnosis are emphasized. This is followed by chapters on the principles and methodology for nucleic acid detection by *in situ* hybridization using radioactive and non-radioactive techniques. The intracellular localization of DNA and messenger RNA receive equal emphasis. There are two chapters on viral gene detection in tissues and cytological smears which are growing areas in laboratory medicine. There is also a chapter on interphase cytogenetics (a relatively new discipline) and its application to unravelling the genetic anomalies which occur in human cancer cells. There is a chapter on nucleolar organizing regions, which have application in tumour biology and prognosis. The final chapter discusses flow cytometry, which gives an overall assessment of total cellular DNA content and phenotyping information on dissociated cells.

Volume II concentrates on the analysis of nucleic acids in extracted material from clinical tissue samples and body fluids. There are three chapters on the principles and protocols for extraction of tissue nucleic acids, preparation and labelling of nucleic acid probes, and protocols for filter hybridization. There is also a comprehensive chapter on DNA sequencing. The polymerase chain reaction (PCR) as it applies to laboratory medicine is described in three chapters; one of these concentrates on the genetic analysis of tissue biopsies and the two others on methodologies for the detection of viruses in tissue and cytological samples. Techniques for detection of other micro-organisms by nucleic acid technology are also described. Protocols and principles of parental identification are given; this is relevant not only in forensic investigations but also in establishing parentage in antenatal counselling in families with genetic disorders. This volume ends with gene screening technologies which have been applied to archival tissue biopsies, and their application to tumour diagnosis and prognosis.

Molecular pathology is a growing discipline and, as our ability to analyse a wider range of clinical samples increases, the precise characterization of molecular abnormalities will assume diagnostic and prognostic importance. *Diagnostic Molecular Pathology: A Practical Approach* aims to provide a synthesis of the molecular techniques available in a form which is useful to diagnostic pathologists. We would like to thank the contributors and Oxford University Press for their help in trying to achieve this aim.

February, 1992

C. Simon Herrington
James O'D. McGee

Contents

Contents

3. Molecular hybridization of nucleic acids   65
Victor T-W. Chan

4. The polymerase chain reaction and the molecular genetic analysis of tissue biopsies   85
Darryl K. Shibata

Contents

Contents

Contents

Contributors

HEIDI M. BAUER
Department of Infectious Diseases, Cetus Corporation, 1400 53rd Street, Emeryville, CA 94608, USA

VICTOR T-W. CHAN
Department of Medicine, Hematology/Oncology Division, Beth Israel Hospital, 330 Brookline Avenue, Boston, MA 02215, USA

JOSEPH A. CHIMERA
Roche Institute of Molecular Biology, 1912 Alexander Drive/7 Southern Kings Parkway, Research Triangle Park, NC 27709, USA

MARCIA EISENBERG
Roche Institute of Molecular Biology, 1912 Alexander Drive/7 Southern Kings Parkway, Research Triangle Park, NC 27709, USA

CATHERINE E. GREER
Department of Infectious Diseases, Cetus Corporation, 1400 53rd Street, Emeryville, CA 94608, USA.

H. ITO
Department of Pathology, Hiroshima University of Medicine, 1-2-3 Kasumi, Minamiku, Hiroshima 734, Japan

MEIRION B. LLEWELYN
MRC Molecular Biology Laboratory, Hills Road, Cambridge CB2 2QH, UK

M. MICHELE MANOS
Department of Infectious Diseases, Cetus Corporation, 1400 53rd Street, Emeryville, CA 94608, USA

BRUCE J. McCREEDY
Roche Institute of Molecular Biology, 1912 Alexander Drive/7 Southern Kings Parkway, Research Triangle Park, NC 27709, USA

CHRIS J. L. M. MEIJER
Department of Pathology, Section of Molecular Pathology, Free University Hospital, De Boelelaan 1117, 1081 HV Amsterdam, The Netherlands

Contributors

PETER W. J. MELKERT
Department of Pathology, Section of Molecular Pathology, Free University Hospital, De Boelelaan 1117, 1081 HV Amsterdam, The Netherlands

DARRYL K. SHIBATA
Department of Pathology, University of Southern California School of Medicine, 1200 N State St 736, Los Angeles, CA 90033, USA

PETER J. F. SNIJDERS
Department of Pathology, Section of Molecular Pathology, Free University Hospital, De Boelelaan 1117, 1081 HV Amsterdam, The Netherlands

JULIA E. STICKLAND
University of Oxford, Nuffield Department of Pathology and Bacteriology, Level 4, Academic Block, John Radcliffe Hospital, Oxford OX3 9DU, UK

E. TAHARA
Department of Pathology, Hiroshima University of Medicine, 1-2-3 Kasumi, Minamiku, Hiroshima 734, Japan

ADRIAAN J. C. VAN DEN BRULE
Department of Pathology, Section of Molecular Pathology, Free University Hospital, De Boelelaan 1117, 1081 HV Amsterdam, The Netherlands

JAN M. M. WALBOOMERS
Department of Pathology, Section of Molecular Pathology, Free University Hospital, De Boelelaan 1117, 1081 HV Amsterdam, The Netherlands

W. YASUI
Department of Pathology, Hiroshima University of Medicine, 1-1-3 Kasumi, Minamiku, Hiroshima 734, Japan

Abbreviations

AMPFLP	amplified fragment length polymorphism
AMPS	ammonium persulfate
AP	alkaline phosphatase
ATP	adenosine triphosphatase
BCIP	5-bromo-4-chloro-3-indolyl phosphate
bio–dNTP	biotin–deoxyribonucleoside triphosphate
bio-dUTP	biotin-deoxyuridine triphosphate
BSA	bovine serum albumin
CSF	cerebrospinal fluid
CsTFA	caesium trifluoroacetate
dATP	deoxyadenosine triphosphate
dCTP	deoxycytidine triphosphate
DEPC	diethyl pyrocarbonate
dGTP	deoxyguanosine triphosphate
dig-dNTP	digoxigenin-deoxyribonucleoside triphosphate
dig–dUTP	digoxigenin–deoxyuridine triphosphate
DMF	dimethyl formamide
DMSO	dimethylsulfoxide
dNTP	deoxyribonucleoside triphosphate
DTT	dithiothreitol
dTTP	deoxythymine triphosphate
dUTP	deoxyuridine triphosphate
EBV	Epstein–Barr virus
EDTA	ethylene diaminetetraacetic acid
ETEC	enterotoxigenic *Escherichia coli*
G–C	guanine–cytosine
HPLC	high performance liquid chromatography
HPVs	human papillomaviruses
ME	mercaptoethanol
MOPS	3-(N-morpholeus)propanesulphonic acid
NBT	nitroblue tetrazolium
NT	nick-translation
OD	optical density
OCT	optimal cutting temperature compound

PAGE	polyacrylamide gel electrophoresis
PBS	phosphate-buffered saline
PCR	polymerase chain reaction
PEG	polyethylene glycol
PETs	paraffin-embedded tissues
PIC	polymorphic information content
PVP	polyvinylpyrrolidone
Rb	retinoblastoma
RF	replicative form
RFLPs	restriction fragment length polymorphisms
RM	reaction mixture
RNasin	ribonuclease inhibitor
rNTPs	ribonucleoside triphosphates
SDS	sodium dodecyl sulfate
SSC	standard saline citrate
SSCP	single-strand conformation polymorphism
SSPE	standard saline phosphate EDTA
SA	streptavidin
TAE	Tris–acetate-EDTA (buffer)
TBE	Tris–borate–EDTA
TCA	trichloroacetic acid
TdT	terminal deoxynucleotidyl transferase
TE	Tris–EDTA (buffer)
TEMED	N',N',N'-tetramethylethylenediamine
TLC	thin-layer chromatography
UV	ultraviolet
VNTRs	variable number of tandem repeats
X-gal	5-bromo-4-chloro-3-indolyl-β-D-galactosidase

Extraction of nucleic acids from clinical samples and cultured cells

VICTOR T-W. CHAN

1. Nucleic acid analysis in the study of human disease

Nucleic acid analysis can identify specific lesions in gene loci, consequent effects on gene expression, and hence their effects on cellular physiology. By using blot hybridization, the presence of abnormal (altered or exogenous) nucleic acid sequences can be identified. The abnormal nucleic acid sequences can then be purified and propagated in large quantity by molecular cloning techniques. These cloned sequences can then be used as the source for characterization of both coding sequences and regulatory regions. Changes in regulatory regions may affect the level of gene expression whilst changes in coding sequences may alter the biological effects of protein products. Cloned sequences can also be used to produce protein products in prokaryotic or eukaryotic expression systems. The functions of these proteins can be studied in a variety of ways to identify their roles in normal cellular physiology and their participation in human disease. The effects of such changes can then be confirmed by introducing the same lesions into endogenous copies of the appropriate gene loci in normal cells by gene targeting in transgenic animals or *in vitro* systems. In general, most studies on human disease follow this line of approach. For example, the introduction of activated oncogenes can induce tumourigenicity in non-tumourigenic cells. On the other hand, introduction of functional tumour suppressor genes into tumour cells can partially suppress their ability to form tumours in animals. The study of gene therapy in genetic diseases by gene targeting is also a major subject. However, the success of such work depends almost entirely on the identification of lesions at specific gene loci which are associated with clinical disease.

1.1 Nucleic acid extraction

It is a necessary prerequisite for the molecular analysis of nucleic acids to obtain them in pure form; that is, free from any contaminants. If clinical samples are to be used, the extraction method should be optimized for the highest yield with high quality since the quantity of clinical sample is usually very limited and often they cannot be easily reobtained. On the other hand, if cultured cells are used, a simpler method may be adequate since the yield can be increased proportionally by scaling up the cultures.

A typical normal diploid human cell contains approximately 7 pg of genomic DNA and a variable amount of mitochondrial DNA. If total DNA is extracted from clinical samples, the end-product is therefore a mixture of genomic and mitochondrial DNA. One way to obtain genomic DNA alone is to isolate nuclei from cells and then extract DNA from the nuclear preparation. However, in most experiments, total DNA can be used as the presence of mitochondrial DNA does not usually cause any problems.

A typical mammalian cell contains approximately 15 pg of RNA. Most of it (80–85%) is ribosomal RNA (28S, 18S, and 5S) with most of the remainder consisting of low molecular weight species (transfer RNAs and small nuclear RNAs). Only 1–5% of the total is messenger RNA (mRNA) which are heterogeneous in both size and sequence. Fortunately, most eukaryotic mRNAs possess polyadenylic acid tails at their 3′ terminus which will anneal to a stretch of oligothymidylic acid, thus allowing the purification of mRNAs by affinity chromatography on oligo(dT)cellulose.

Several methods for DNA and RNA extraction are described in this chapter. These methods are easy to follow and to perform, even for relatively inexperienced workers. Only nucleic acid extraction from cellular material and fresh/frozen tissue is described. A protocol for the extraction of DNA from archival biopsies is given in Chapter 10. The precise method used should be determined by the requirements of the experiment.

2. Extraction of DNA

Genomic DNA is usually isolated by disrupting the cytoplasmic and nuclear membranes of the cells followed by the removal of cellular proteins. Sample preparation is described in *Protocol 1*. There are two ways to achieve subsequent deproteinization:

- digestion using proteases (proteinase K for example) in the presence of EDTA and detergent (such as sodium sarcosine)
- denaturation and extraction by organic solvent (for example, phenol and chloroform)

These methods are described in *Protocols 2* and *3* respectively.

Genomic DNA can be precipitated from aqueous solution by the addition of absolute ethanol or isopropanol. The end-products of these two methods are DNA fragments of size 100–200 kb which are generally suitable for

Southern blot analysis and molecular cloning with bacteriophage lambda. However, larger DNA fragments are required for molecular cloning into cosmid vectors (250–300 kb). Therefore, one should always try to obtain fragments of the largest possible size in a DNA preparation. The two main factors affecting the final size of DNA molecules are cellular nucleases and mechanical shearing during extraction. Endogenous nucleases should be inactivated immediately when the cell is lysed. This requires thorough mixing of the cell lysate with nuclease inactivators. However, if mixing is too vigorous, DNA molecules will be sheared mechanically. These two factors should be balanced and samples should therefore be mixed thoroughly but gently. Furthermore, mechanical shearing of DNA can also be produced during extraction with organic solvents so care should be taken during these steps.

2.1 Extraction of DNA using proteinase K

The procedure described in *Protocols 1* and *2* is a modification of that of Blin and Stafford (1). The cells are lysed by sodium sarcosine and cellular proteins digested using proteinase K in the presence of EDTA.

Protocol 1. Preparation of samples

Materials
- Extraction buffer: 300 mM sodium acetate, 50 mM EDTA, pH 7.5
  Add RNase A to a final concentration of 25 µg/ml before use.
- RNase A: dissolve RNase to 20 mg/ml in sterile distilled water, boil the solution for 10 min (to inactivate DNase) and chill on ice for 5 min. Store at $-20°C$.
- Phosphate-buffered saline (PBS): 10 mM phosphate, 150 mM NaCl, pH 7.2
- Sodium sarcosine: dissolve sodium sarcosine to 10% (w/v) in sterile distilled water and heat the solution to 65°C for at least 1 h.

Method
The precise method used depends upon the type of sample.

A. Cultured cells
1. Collect the cells ($1-5 \times 10^7$) by centrifugation in a 50-ml polypropylene tube at 800 g for 7 min at 4°C. Monolayer cells require detachment by trypsinization prior to centrifugation (see Volume I) while cells in suspension can be collected directly.
2. Wash the cells in 20 ml ice-cold PBS.

3

Protocol 1. *Continued*

3. Resuspend the cells in 10 ml extraction buffer. Alternatively, they can be resuspended in 1 ml extraction buffer, quickly frozen in liquid nitrogen and stored at −70°C or below. DNA can be extracted by adding 9 ml of pre-warmed extraction buffer containing 110 μg/ml proteinase K and 0.55% (w/v) sodium sarcosine.

B. Tissue samples

1. Tissue samples for nucleic acid extraction should always be stored at −70°C or below.

2. Pre-cool the stainless-steel cup of a Waring blender (Patterson Scientific) on dry ice.

3. Grind approx. 5 g of dry ice to a fine powder.

4. Add the frozen tissue to the powder in the blender cup and grind.

5. Add additional dry ice during grinding to ensure that the tissue remains frozen.

6. Transfer the ground tissue/dry ice mixture to a 50-ml polypropylene tube, and let most (but not all) of the dry ice evaporate.

7. Add 10 ml extraction buffer per gram of tissue.

C. Blood

1. Transfer 20 ml of EDTA-anticoagulated blood to a 50 ml polypropylene tube containing 30 ml 37.5 mM NaCl, pH 7.5 and mix gently.

2. Collect the white cells by centrifugation at 800 g for 7 min at 4°C.

3. Discard the supernatant (lysed red cells) and repeat the washing once with the same solution.

4. Resuspend the cells in 10 ml extraction buffer. Alternatively, the cells can be resuspended in 1 ml extraction buffer, quickly frozen in liquid nitrogen and stored at −70°C (or below). DNA can be extracted by adding 9 ml pre-warmed extraction buffer containing 110 μg/ml proteinase K and 0.55% sodium sarcosine.

Protocol 2. DNA extraction by proteinase K[a]

Materials

(see also *Protocol 1*)

- Proteinase K
  Dissolve lyophilized proteinase K to 20 mg/ml in sterile distilled water and store at −20°C in small aliquots.

- Buffer-saturated phenol
  Most high-grade phenol can be used directly without further distillation.
  Melt phenol at 65 °C in the presence of an equal volume of 0.5 M Tris–HCl,
  pH 8.0 containing 0.2% 8-hydroxyquinoline and 0.2% 2-mercaptoethanol
  (2-ME). Once the phenol is completely melted, mix thoroughly, discard the
  aqueous phase and extract twice with 0.1 M Tris–HCl pH 8.0 containing
  0.2% 2-ME. After the final extraction, leave the aqueous phase in the
  bottle. In this way, phenol can be stored for up to 1 month at 4 °C.

- Chloroform: isoamylalcohol (24:1, v/v) mixture.

- TE buffer: 10 mM Tris–HCl, 1 mM EDTA, pH 8.0.

Method

1. Prepare samples as described in *Protocol 1*.

2. Add 50 μl proteinase K stock solution to each sample and mix gently.

3. Add 250 μl sodium sarcosine solution to each sample, and mix gently and
   thoroughly.

4. Spin the samples for 30 sec at 3500 g.

5. Incubate the samples at 50 °C for 5 h. For convenience, the incubation
   can be prolonged to 16 h (overnight).

6. Add 10 ml buffer-saturated phenol to each sample, and mix until a
   homogeneous emulsion is formed.

7. Allow extraction to occur by mixing for 5 min at room temperature.

8. Spin for 1 min at 3500 g.

9. Add 10 ml chloroform: isoamylalcohol mixture and repeat the extraction
   for 5 min.

10. Separate the aqueous phase from the organic phase by centrifugation at
    3500 g for 6 min at room temperature.

11. Transfer the viscous aqueous phase to a fresh tube with a wide-bore
    pipette and repeat the extraction with chloroform: isoamylalcohol mix-
    ture once or twice (until the interface is clear).

12. Add 2 vol. absolute ethanol to the aqueous phase and mix thoroughly
    and gently. The DNA should precipitate immediately.

13. Spin down the DNA at 3500 g for 5 min.

14. Discard the supernatant and wash the DNA pellet with 20 ml 70%
    ethanol.

15. Wash the DNA pellet with 1 ml 70% ethanol and transfer to microfuge
    tubes.

16. Spin down the DNA for 1 min at 3500 g and discard the supernatant.

Protocol 2. *Continued*

17. Dry the DNA pellets under vacuum for 2.5 min.

18. Dissolve the DNA in TE buffer at approx. 1 mg/ml (see Section 6.2).

[a] Although this protocol may take 1–2 days to carry out, it is not particularly labour-intensive. This is because, after proteinase K digestion, most of the cellular proteins are digested into oligopeptides, which give only a small interface during phenol/chloroform extraction, making the procedure technically easier.

2.2 DNA extraction by a rapid phenol method

Protocol 3 is adapted from an original method which used frozen phenol (2). The cells are lysed by sodium sarcosine and hot phenol which also extracts the cellular proteins.

Protocol 3. DNA extraction by rapid phenol method[a]

Materials

(see also *Protocols 1* and *2*)

- Extraction buffer (without RNase)
- 10% (w/v) sodium sarcosine
- Buffer-saturated phenol
- Chloroform:isoamylalcohol (24:1, v/v) mixture

Method

1. Prepare samples for DNA extraction according to *Protocol 1*, but resuspend the cells (or tissue samples) in 2.5 ml of extraction buffer (without RNase A).

2. During sample preparation, pre-heat a mixture of 7.5 ml extraction buffer containing 0.5% sodium sarcosine, and 10 ml buffer-saturated phenol to 65°C.

3. Pour the hot phenol/buffer quickly into the samples and mix thoroughly but gently until a homogeneous emulsion is formed.

4. Allow extraction to occur by mixing at room temperature for 5 min.

5. Spin the samples briefly at 3500 g.

6. Add 10 ml of chloroform:isoamylalcohol mixture and mix at room temperature for 5 min.

7. Separate the aqueous phase from the organic phase by centrifugation at 3500 g for 6 min at room temperature.

8. Transfer the aqueous phase and the protein interface to a fresh tube with a wide-bore pipette.

9. Add 10 ml chloroform:isoamylalcohol mixture and mix at room temperature for 5 min.

10. Separate the two phases by centrifugation at 3500 g for 5 min and transfer only the aqueous phase to a fresh tube.

11. Add 10 ml chloroform:isoamylalcohol mixture, mix at room temperature for 5 min, separate the aqueous and organic phases as in step 10 and transfer the aqueous phase to a fresh tube.

12. Repeat step 11, then precipitate the DNA by adding 2 vol. of absolute ethanol to the aqueous phase.

13. Mix the samples gently and thoroughly. The DNA should precipitate immediately.

14. Spin down the DNA at 3500 g for 5 min, discard the supernatant and wash the DNA pellets with 20 ml 70% ethanol.

15. Wash the DNA pellet in 1 ml 70% ethanol and transfer to 1.5 ml microfuge tubes.

16. Spin down the DNA at 3500 g for 1 min and discard the supernatant.

17. Dry the pellets under vacuum for 2.5 min.

18. Dissolve the DNA in TE buffer (containing 2.5 µg/ml RNase A) at approx. 1 mg/ml (see Section 6.2).

---

[a] The rapid phenol method is considerably quicker than the proteinase K method. However, it is slightly more labour-intensive because all the cellular proteins have to be extracted by phenol/chloroform. There may still be trace amounts of protein in the final product but this does not usually interfere with subsequent procedures (for example, gel electrophoresis or restriction digestion).

2.3 Isolation of high-molecular-weight DNA

The procedure described in *Protocol 4* is a modification of that of Kupiec *et al.* (3). The cells are lysed and the cellular proteins digested as in the proteinase K method (see *Protocol 2*). However, this method avoids phenol/chloroform extraction and thus minimizes mechanical shearing of DNA.

Protocol 4. Isolation of high-molecular-weight DNA

Materials
(see also *Protocols 1* and *2*)

• Extraction buffer
• RNase A, 20 mg/ml
• Proteinase K, 20 mg/ml

Protocol 4. *Continued*

- 10% (w/v) sodium sarcosine
- Denaturing buffer: 80% formamide, 20 mM Tris–HCl, pH 8.0.
  Most high-grade formamide can be used directly without any pre-treatment.
  However, if a yellow colour is observed, deionize the formamide by mixing
  with a mixed bed resin (for example, Bio-Rad AG 501-XG) on a shaker for
  30 min and separate it from the resin either by centrifugation at 400 *g* for 5 min
  or filtration. Store the deionized formamide at −70°C in small aliquots.
- Dialysis buffer: 0.1 M sodium acetate, 10 mM Tris–HCl, pH 8.0, 10 mM
  EDTA
- TE buffer

Method

1. Prepare samples for DNA extraction and set up the proteinase K digestion
   according to *Protocol 1* and steps 1–5 of *Protocol 2*.
2. Add 10 ml of denaturing buffer pre-cooled to 4°C to each sample and mix
   very gently.
3. Incubate the samples at 15°C for 6 h.
4. Pour the viscous solution into collodion bags (Sartorius SM 13200E or
   equivalent) and dialyse three times against 5 litres of dialysis buffer.
5. Dialyse the solution five times against 5 litres of TE buffer.
6. The DNA solution may be concentrated by placing the collodion bag in
   direct contact with dry resin (for example, Sephadex G-50) (optional).
7. Transfer the viscous DNA solution to a microfuge tube(s).[a]

---

[a] DNA extracted by this method is 200–300 kb in length and is therefore good starting material
for molecular cloning with cosmid vectors.

---

3. Extraction of RNA

The basic principle of RNA extraction is very similar to that for DNA but the
methods used are different in many specific aspects. In general, the cells are
lysed, the cell lysate de-proteinized and the RNA separated from DNA.
However, RNA is much more labile than DNA and more susceptible to
nuclease degradation. Furthermore, RNases are much more resistant to
protein denaturants than DNases. These factors make the isolation of un-
degraded RNA relatively difficult. Therefore, in order to obtain good quality
RNA, it is necessary to lyse the cells and inactivate RNases simultaneously.
Since RNA molecules are generally quite small, the cell lysate can be mixed
vigorously with RNase inactivators or inhibitors without undue worry about

mechanical shearing. Care should be taken to avoid the accidental introduction of RNases from other sources. Since there is a significant amount of RNase on the skin, disposable gloves should be worn during the isolation of RNA. Furthermore, especially for inexperienced workers, the following precautions should be taken:

- All water used should be treated with diethylprocarbonate (DEPC), 0.1% at 37°C overnight, then autoclaved. After this treatment, the water is essentially free of RNase and can be used to prepare solutions or to rinse any items used for RNA isolation. Note: DEPC is a suspected carcinogen and should be handled with great care.

- Sterile, disposable plasticware should preferably be used because it is RNase-free. If general laboratory glassware or plasticware is used, it should be pre-soaked in DEPC-treated water for 2 h at 37°C. The DEPC-treated items should be rinsed thoroughly with DEPC-treated water and then autoclaved.

- Solutions for RNA extraction should be treated with 0.1% DEPC at 37°C overnight and then autoclaved, or be made up using DEPC-treated water. Note, however, that DEPC reacts quickly with amines and so cannot be used to treat solutions containing Tris. Tris buffers should be made using molecular biology grade Tris and DEPC-treated water, then autoclaved.

- It is a good idea to reserve such items as glassware, plasticware, micropipettes, electrophoresis tanks, gel trays, etc., for work involving RNA and to designate an area where RNase must not be used.

- Wherever possible, RNase inhibitors should be used in the extraction solutions, the subsequent steps in the experiment, and for short-term storage of RNA in aqueous solution. Protein inhibitors of RNase are commercially available from several manufacturers (for example, Sigma) and are generally ready to use directly. They should be stored and used as recommended by the suppliers. Since these inhibitors can be denatured by protein denaturants or be degraded by proteases, they are relatively ineffective when these agents are used.

- Vanadyl–ribonucleoside complexes are also commercially available (for example, Sigma). The complexes formed between the oxovanadium ions and the four ribonucleosides are transition-state analogues that bind to many RNases and almost completely inhibit their activity. The complexes are added to the solutions for extraction and all subsequent steps of the procedure. However, vanadyl–ribonucleoside complexes strongly inhibit the cell-free translation of RNA and must be removed from a mRNA preparation by multiple phenol extractions if it is to be analysed in this way.

3.1 Extraction of RNA using proteinase K

The method given in *Protocol 5* is a modification of the procedure of Favaloro *et al.* (4) and can only be used with samples that can be dispersed readily into single cells. Thus, it is not suitable for RNA extraction from frozen tissue because sodium dodecyl sulfate (SDS) and proteinase K cannot gain access to the inner part of the tissue. Moreover, endogenous RNases can cause significant degradation of RNA before it is digested by proteinase K.

Protocol 5. RNA extraction using proteinase K

Materials
(see also *Protocols 1* and *2*)
All reagents should be RNase-free or DEPC-treated.
- Extraction buffer: 0.15 M NaCl, 10 mM Tris–HCl, pH 8.5, 0.5% Nonidet P-40, 1 mM dithiothreitol (DDT), 20 mM vanadyl–ribonucleoside complexes
- Proteinase K buffer: 0.2 M Tris–acetate, 0.3 M sodium acetate, 50 mM EDTA, pH 7.5, 2% (w/v) SDS
- SDS solution: dissolve SDS to 10% (w/v) in sterile distilled water and treat the resulting solution at 65°C for at least 1 h
- Proteinase K solution, 20 mg/ml
- Buffer-saturated phenol
- Chloroform: isoamylalcohol (24:1, v/v) mixture
- TE buffer

Method
1. Prepare the samples for RNA extraction as described in *Protocol 1*.
2. After washing in PBS, re-suspend the cells in 1 ml of PBS and transfer the cell suspension to a microfuge tube.
3. Spin down the cells at 14 000 g for 7 sec at 4°C.
4. Re-suspend the cells in 200 μl of ice-cold extraction buffer by vortexing.
5. For extraction of cytoplasmic RNA, centrifuge the cell lysate at 14 000 g for 2 min at 4°C. Transfer the supernatant to a fresh tube. The pellet, which contains the nuclei, should be discarded.
6. Add 200 μl of proteinase K buffer, containing 400 μg/ml proteinase K, to each sample.
7. Incubate samples at 37°C for 30 min.[a]
8. Extract the sample by adding an equal volume of buffer-saturated phenol:chloroform:isoamylalcohol mixture (1:1, v/v)[b] and mixing the mixture at room temperature for 5 min.
9. Centrifuge the sample at 14 000 g for 2 min at room temperature.
10. Transfer the aqueous phase to a fresh tube and extract with an equal volume of chloroform:isoamylalcohol mixture for 5 min at room temperature.

10

11. Repeat step 10 once.

12. Transfer the aqueous phase to a fresh tube and add 2.5 vol. of absolute ethanol to the sample.

13. Precipitate the RNA on dry ice for 30 min.

14. Centrifuge at 14 000 g for 10 min at 4°C to pellet the RNA and wash each pellet in 700 μl of 80% ethanol.

15. Dry the RNA pellets under vacuum for 2.5 min, dissolve in 300 μl of TE buffer and add 750 μl absolute ethanol.

16. Store the samples at −70°C.

17. To recover the RNA, transfer an aliquot to a microfuge tube and add 3 M sodium acetate to give a final concentration of 0.3 M.

18. Mix thoroughly and incubate the solution at −70°C for 15 min.

19. Pellet the RNA by centrifugation at 14 000 g for 10 min at 4°C.

20. Wash the pellet in 80% ethanol, dry it under vacuum and dissolve the RNA in TE buffer.

[a] For extraction of total RNA, step 5 should be avoided. After the addition of proteinase K, draw the lysate into a hypodermic syringe fitted with a 23-gauge needle and expel it back into the tube. Repeat five times. Following proteinase K digestion, add 200 μl 3 M sodium acetate, pH 5.2 and 600 μl water-saturated phenol (instead of buffer-saturated) to the lysate. Extract by mixing the sample at room temperature for 5 min. Centrifuge the sample at 14 000 g for 2 min. Transfer the aqueous phase to a fresh tube and continue the following steps.
[b] This gives proportions of sample:buffer-saturated phenol:chloroform:isoamylalcohol of 50:25:24:1.

3.2 Extraction of RNA using guanidinium thiocyanate

The procedure given in *Protocol 6* is a modification of those of Glisin *et al*. (5) and Chirgwin *et al*. (6). It has been used successfully on samples that are particularly rich in RNases and is probably the most widely used method for RNA extraction. Guanidinium thiocyanate is a very strong protein denaturant, which is used to lyse the cells and denature all the cellular proteins (including RNases) simultaneously. Since the buoyant density of RNA (approx. 1.9 g/ ml) is much higher than that of DNA and proteins, it can be separated from other cellular components by equilibrium density gradient centrifugation on caesium chloride (CsCl).

Protocol 6. RNA extraction by guanidinium thiocyanate

Materials

All reagents should be RNase-free or DEPC-treated.

- Homogenization buffer: 4 M guanidinium thiocyanate, 100 mM Tris–HCl, pH 7.5

  Add 2-mercaptoethanol to 1% (v/v) before use.

Protocol 6. *Continued*

- Caesium chloride: 5.7 M in 10 × TE buffer
- 10 × TE buffer: 100 mM Tris–HCl, 10 mM EDTA, pH 8.0

Method

1. Prepare the samples for RNA extraction as described in *Protocol 1*.

2. Add 3 ml of homogenization buffer to each sample and mix vigorously.

3. Draw the homogenate into a sterile hypodermic syringe fitted with a 23-gauge needle and expel it back into the tube.

4. Repeat step 3 five times.

5. Spin down any lumps of un-homogenized tissue at 3500 *g* for 10 min at room temperature (this can be omitted if single-cell suspensions are used).

6. Two density gradient centrifugation approaches are possible. For example, for a Beckman SW41 rotor (or its equivalent), add 8 ml of 5.7 M CsCl solution to a clear centrifuge tube, and layer the homogenate on to the CsCl cushion. Alternatively, add 0.8 g of CsCl per millilitre to the homogenate and layer the resulting solution onto 1.5 ml of CsCl cushion in a clear centrifuge tube (for example, when using the Beckman SW55 rotor). The DNA will band at the top of the CsCl cushion whilst the RNA will form a pellet at the bottom of the tube.[a] Although the RNA is intact, the DNA is too small (approx. 30 kb) to be useful in many experiments.

7. Centrifuge at 200 000 *g* for 16 h at 20°C (for example, 45 000 r.p.m. when using a SW55 rotor) or at 135 000 *g* for 24 h at 20°C (for example, 32 000 r.p.m. when using a SW41 rotor).

8. Remove the solution above the cushion. The cellular DNA usually forms a visible white band at the top of the CsCl cushion and can be removed at this stage.

9. Remove the remaining solution with a micropipette (be careful not to disturb the RNA pellet).

10. Add 750 μl of 80% ethanol to the tube and transfer the RNA pellet with the ethanol to a microfuge tube.

11. Wash the original centrifuge tube with 750 μl of 80% ethanol and combine the ethanol washes. The RNA pellet can be stored at −70°C.

12. Spin down the RNA at 14 000 *g* for 5 min at room temperature.

13. Discard the supernatant, wash the pellet in 1 ml of 80% ethanol and spin at 14 000 *g* for 5 min at room temperature.

14. Discard the supernatant and dry the pellet under vacuum for 2.5 min.

15. Dissolve the RNA pellet in 300 μl of TE, add 750 μl of absolute ethanol and store the samples at −70°C.

16. To recover the RNA by centrifugation, follow steps 17–20 of *Protocol 5*.

---

a Alternatively, add 0.75 ml of caesium trifluoroacetate (CsTFA: density = 2 g/ml) to a centrifuge tube followed by a cushion of CsTFA at a density of 1.7 g/ml. Adjust the density of the homogenate to 1.4 g/ml by adding CsTFA, then layer this on top of the CsTFA cushion. After centrifugation, the RNA and DNA bands are at the top of the first (bottom) and second cushions respectively. The addition of ethidium bromide to a final concentration of 25 μg/ml will enhance visualization of nucleic acids and the DNA and RNA can be recovered by side-puncture of the centrifuge tube with a syringe and needle. Ethidium bromide can be removed prior to nucleic acid precipitation by butan-1-ol extraction (see *Protocol 3*, Chapter 2).

---

4. Simultaneous extraction of DNA and RNA

Extraction of high-molecular-weight DNA and RNA is a prerequisite for many studies employing molecular biology techniques. Efficient extraction of DNA and RNA may not be important when experimentally-derived samples (for example, cultured cells) are used because the nucleic acid yield can be increased simply by scaling up the experimental procedure. However, it is crucial for studies on clinically-derived samples because biopsies taken primarily to establish a tissue diagnosis are frequently small. Furthermore, biopsy samples from patients cannot be re-obtained easily. One way to overcome this problem is to extract DNA and RNA simultaneously with high quality and high yield. However, extraction of high-molecular-weight DNA generally requires gentle tissue disruption, which hinders access of nuclease inactivators to the cells and allows extensive degradation particularly of RNA, while vigorous cell disruption leads to considerable shearing of DNA. To minimize the degradation of RNA while still obtaining high-molecular-weight DNA, grinding frozen tissue in the presence of frozen phenol and allowing them to melt simultaneously produces intimate contact of nucleases and nuclease inactivators. This facilitates optimal extraction of high-molecular-weight DNA and undegraded RNA from very limited material with a high yield (2). This procedure is detailed in *Protocol 7*.

Protocol 7. The frozen phenol method for simultaneous extraction of high-molecular-weight DNA and RNA

Materials
(see also *Protocols 1* and *2*)
All reagents should be RNase-free or DEPC-treated.
- Extraction buffer: 300 mM sodium acetate, 50 mM EDTA, pH 7.5. Add 0.5% (w/v) sodium sarcosine before use.
- Sodium sarcosine: 10% (w/v) in sterile distilled water, heat-treated at 65°C for at least 1 h

Protocol 7. *Continued*

- Buffer-saturated phenol
- Chloroform:isoamylalcohol (24:1, v/v) mixture
- 10 × TE buffer: 100 mM Tris–HCl, 10 mM EDTA, pH 8.0
- Caesium chloride (CsCl) solution: 1.3 g CsCl per millilitre of 10 × TE (final density = 1.7 g/ml)
- Ethidium bromide, 10 mg/ml in sterile distilled water
- 3 M sodium acetate, pH 5.2

Method

1. Mix 5 ml of buffer-saturated phenol and 5 ml of extraction buffer in a 50-ml polypropylene tube.
2. Homogenize the mixture by vortexing, and immerse the homogeneous emulsion quickly in liquid nitrogen until it starts to freeze.
3. Pre-cool the stainless steel cup of a Waring blender (Patterson Scientific) on dry ice.
4. Grind approx. 5 g of dry ice to a fine powder and add a small aliquot of frozen phenol/extraction buffer.
5. Grind the frozen phenol/extraction buffer to a fine powder and add more frozen phenol/extraction buffer to the blender cup.
6. Repeat the grinding until the whole frozen phenol/extraction buffer mixture is ground to a fine powder.
7. Add the frozen tissue to the phenol/extraction buffer powder and grind the tissue to a fine powder. Dry ice should be added during the grinding to ensure that the tissue and phenol/extraction buffer are completely frozen.
8. Transfer the whole mixture to a 50-ml polypropylene tube and melt the phenol/extraction buffer/tissue mixture quickly in a 65°C water bath with gentle agitation. Do not put the cap on tightly.
9. When the phenol begins to melt, shake the tube gently and thoroughly to ensure that the whole mixture forms a homogeneous emulsion.[a]
10. When the phenol is completely melted, mix the mixture for 5 min at room temperature.
11. Spin the sample for 1 min at 3500 g and add 5 ml of chloroform: isoamylalcohol mixture.
12. Mix the sample in the resultant phenol/chloroform:isoamylalcohol mixture for 5 min at room temperature.
13. Separate the aqueous and organic phases by centrifugation at 3500 g for 5 min at room temperature.

14. Transfer the aqueous phase and protein interface to a fresh tube, add an equal volume of chloroform:isoamylalcohol mixture and mix for 5 min at room temperature.

15. Separate the two phases by centrifugation at 3500 g for 5 min at room temperature and transfer only the aqueous phase to a fresh tube.

16. Repeat step 15 once.

17. Add 0.8 g caesium chloride to each millilitre of tissue extract.

18. Add 3 ml of CsCl solution to a centrifuge tube and layer the CsCl solution of tissue extract from step 17 carefully onto this CsCl cushion. The addition of ethidium bromide (25 µg/ml) to the tissue extract will enhance the subsequent visualization of nucleic acids.

19. Centrifuge the samples at 135 000 g for 24 h at 25°C (for example, 32 000 r.p.m. in a Beckman SW41 rotor).

20. After centrifugation, the DNA should form a band at the top of the CsCl cushion whilst the RNA forms a pellet at the bottom of the tube.

21. Recover the DNA with a wide-bore pipette and transfer to a fresh tube. Remove the remaining solution with a micropipette (be careful not to disturb the RNA pellet).

22. Add 750 µl of 80% ethanol to the tube and transfer the RNA pellet with the ethanol to a microfuge tube.

23. Wash the centrifuge tube with 750 µl of 80% ethanol and combine the ethanol washes.[b]

24. Add an equal volume of sterile distilled water to the DNA solution then 2 × the original (1 × the final) volume of isopropanol. Mix gently to precipitate the DNA.

25. Spin the DNA down at 3500 g for 5 min at room temperature.

26. Discard the supernatant and wash the DNA pellet 2–3 times in 70% ethanol.

27. Dry the DNA pellet under vacuum for 2.5 min and dissolve the DNA in TE buffer at approx. 1 mg/ml (see Section 6.2).

28. Spin down the RNA at 14 000 g for 5 min at room temperature and discard the supernatant. Wash the pellet in 80% ethanol.

29. Dissolve the RNA pellet in 300 µl of TE buffer, add 750 µl of absolute ethanol and store the samples at −70°C.

30. To recover the RNA by centrifugation, follow steps 17–20 of *Protocol 5*.

---

[a] The DNA and RNA extracted with this method are of high quality, suitable for Southern and Northern analysis. However, rapid inactivation of nucleases is crucial to the success of the method. Therefore, the samples must be mixed thoroughly to form a homogeneous emulsion when the phenol is melted.

[b] The DNA and RNA can be stored at this stage at 4°C and −70°C respectively. Ethidium bromide can be removed by extraction with butan-1-ol (see *Protocol 3*, Chapter 2).

5. Purification of poly (A)$^+$ RNA

Since the poly (A)$^+$ tail of poly (A)$^+$ RNA can anneal to oligodeoxythymidylic acid [oligo(dT)] in high salt buffers, and, since oligo(dT) can be attached covalently to solid matrices (for example, cellulose), this allows the purification of mRNA from a complex population of cellular RNA. When total RNA, in high salt buffer, passes down a column of oligo(dT) cellulose, only poly (A)$^+$ RNA can anneal and hence be immobilized. All other RNAs are washed off. The poly (A)$^+$ RNA is then eluted by low salt buffer or water and recovered by ethanol precipitation. This procedure is described in *Protocol 8*.

Protocol 8. Isolation of poly (A)$^+$ RNA

Materials

All reagents should be RNase-free or DEPC-treated.

- 2 × loading buffer: 50 mM Tris–HCl, pH 7.5, 1 M NaCl, 5 mM EDTA, 0.2% (w/v) SDS
- Oligo(dT) cellulose
- 5 M NaCl

Method

1. Re-suspend 1 g of oligo(dT) cellulose in sterile distilled DEPC-treated water and pour a column of approx. 1 ml packed volume in a disposable column or syringe.
2. Wash the column with hot (65°C) water followed by 5 ml of 1 × loading buffer.
3. Dissolve the RNA sample in 250 μl water and heat the solution at 65°C for 5 min. Cool the sample rapidly to room temperature and add an equal volume of 2 × loading buffer.
4. Load the sample on to the column and collect the eluate immediately.
5. Wash the column with 1 ml of loading buffer and combine the eluates.
6. Heat the eluate at 65°C for 5 min, cool rapidly to 20°C and re-apply it to the column.
7. Wash the column with 10 × 1 ml of loading buffer to elute all poly (A)$^-$ RNA.
8. Elute the poly (A)$^+$ RNA with 1.8 ml of hot (65°C) water and collect the eluate immediately.
9. Heat the eluate at 65°C for 5 min and cool it rapidly to room temperature.
10. Add 0.2 ml of 5 M NaCl, load the sample on to a new column, and repeat steps 4–7.

11. Elute the poly (A)$^+$ RNA with two 1-ml fractions of hot (65°C) water.

12. Concentrate the eluate to approx. 400 μl by butan-1-ol extraction.

13. Add 2.5 vol. of absolute ethanol to the sample and store at −70°C.

14. Poly (A)$^+$ RNA can be recovered by following steps 17–20 of *Protocol 5*.

6. Analysis of DNA and RNA

After DNA and/or RNA has been isolated, its purity and size should be determined prior to further analysis. Impurities in nucleic acid preparations can inhibit subsequent enzyme reactions, particularly restriction digestion of DNA. If this occurs, proteinase K digestion and phenol/chloroform extraction should be repeated: this usually solves the problem. If RNA samples are to be used solely for Northern blot analysis (see Chapter 3), enzymatic manipulation is not necessary. However, DNA samples should be of sufficient size for the purpose of the investigation, and RNA samples should be undegraded.

6.1 Qualitative analysis of nucleic acids

DNA or RNA (single-stranded or double-stranded) migrate through gel matrices at a rate inversely proportional to the $\log_{10}$ of their molecular weights. However, single-stranded molecules migrate more quickly than double-stranded molecules of the same length. Since the molecular weight of nucleic acids is proportional to their size (length in base-pairs or kilobase-pairs), a standard curve can be constructed by plotting $\log_{10}$ of the size of the marker fragments against the distance they migrate. The resulting line can be used to calculate the size of a particular sample fragment from the distance it has migrated.

6.1.1 Gel electrophoresis of DNA

DNA fragments are usually fractionated using agarose gels in Tris–borate EDTA (TBE) buffer. A general procedure is given in *Protocol 9*.

Protocol 9. Agarose gel electrophoresis

Materials

- 10 × Tris–borate–EDTA (TBE): 0.9 M Tris base, 0.9 M boric acid, 20 mM EDTA
  Adjust the pH to 8.1–8.2 using solid boric acid.
- Agarose (electrophoresis grade)
- Ethidium bromide, 10 mg/ml in sterile distilled water
- 6 × loading buffer: 0.25% bromophenol blue, 40% (w/v) sucrose in 1 × TBE

Protocol 9. *Continued*

Method

1. Prepare a sufficient volume of 1 × TBE to fill the electrophoresis tank and prepare the gel.

2. Add sufficient agarose to 1 × TBE buffer to give 0.35–0.45% (w/v) and heat in a microwave oven until the agarose dissolves.

3. Cool the solution to approx. 50°C. While cooling occurs, seal the edges of a clean UV-transparent gel tray with autoclave tape.

4. Add ethidium bromide to the agarose solution to a final concentration of 0.5 μg/ml and mix thoroughly. Avoid producing any air-bubbles in the gel solution.

5. Pour and evenly distribute the warm agarose solution into the gel tray and insert the comb vertically with its teeth approx. 1.5 mm above the bottom of the agarose solution.

6. Let the agarose gel set for 30 min, then carefully remove the gel comb and autoclave tape. Place the gel tray (containing the agarose gel) into an electrophoresis tank containing a sufficient amount of 1 × TBE buffer to which ethidium bromide has been added to a final concentration of 0.5 μg/ml.

7. Prepare the DNA samples and molecular weight markers by mixing them with loading buffer (5:1, v/v). 200 ng of marker DNA per 5 mm well is sufficient to give a strong signal by ethidium bromide staining. Marker fragments of size appropriate to the experiment should be used to construct a standard curve.

8. Load the DNA samples and markers carefully into the wells. Marker fragments should be loaded on either side of the test samples to facilitate sample size determination.[a]

9. Run the gel at a voltage gradient of 1–3 V/cm of gel for 1–3 h.[b]

10. Examine the gel on a UV transilluminator and photograph it with a Polaroid camera.

---

[a] For Southern analysis and genomic cloning with bacteriophage lambda, the DNA should be approx. 200 kb in size, whereas for cosmid cloning it should be approx. 300 kb.
[b] Until the bromophenol blue approaches the bottom of the gel.

6.1.2 Gel electrophoresis of RNA

RNA samples can be preliminarily analysed by agarose gel electrophoresis in TBE buffer (see *Protocol 9*) to test if the samples are significantly degraded. For a good RNA preparation, the intensity of the 28S RNA band should be approximately three times that of the 18S band on a TBE gel. A lower ratio of intensity of these two bands indicates degradation. However, for more precise characterization, RNA samples should be analysed by agarose gel electrophoresis after denaturation, or on a denaturing gel (8, 9). These two approaches are described in *Protocols 10* and *11* respectively.

Protocol 10. Electrophoresis of RNA denatured with glyoxal and DMSO

Materials

All reagents should be RNase-free or DEPC-treated.

- 6 M glyoxal:glyoxal is usually obtained as a 40% solution (6 M). Since it oxidizes rapidly in air, glyoxal solution must be deionized before use by mixing with a mixed bed resin (for example, Bio-Rad AG 501-X8) on a rotary shaker until the pH is above 5.0. Deionized glyoxal should be stored at −20°C in small aliquots in tightly capped tubes: the tubes should be almost completely filled and each aliquot should be used only once.
- Dimethylsulfoxide (DMSO): In general, most high-grade DMSO can be used directly without any pre-treatment. However, a new bottle of DMSO should preferably be aliquoted and stored at −20°C. Each aliquot should be used only once.
- 10 × electrophoresis buffer: 100 mM sodium phosphate, pH 7.0
- Denaturing mixture: 3 ml of 6 M glyoxal, 7.5 ml of DMSO and 1.5 ml of 10 × electrophoresis buffer. This should be stored at −20°C in small aliquots in tightly capped tubes. Fill the tubes almost completely and use only once.
- Loading buffer: 50% glycerol, 0.25% bromophenol blue in 1 × electrophoresis buffer
- Agarose (electrophoresis grade)

Method

1. Dissolve each RNA sample (containing up to 20 μg RNA) in 3 μl of DEPC-treated water and add 12 μl of denaturing mixture.
2. Denature the RNA by heating at 50°C for 1 h.[a]
3. Prepare a 1.1% (w/v) agarose gel in 1 × electrophoresis buffer (see *Protocol 9*) and cool the solution to 70°C. Add solid sodium iodoacetate to a final concentration of 10 mM to inactivate RNases.
4. Cool the mixture to approx. 50°C, pour it into a gel tray and allow it to set at room temperature for 30 min.
5. Add 2.5 μl of loading buffer to each sample, mix the reagents gently and spin the samples by pulsing to 12 000 g in a microfuge.
6. Load the samples into the wells of the agarose gel and run it at 1–2 V/cm in 1 × electrophoresis buffer. During electrophoresis, the gel running

Protocol 10. *Continued*

buffer must be recirculated in order to keep its pH at approx. 7.0 as glyoxal dissociates from RNA at pH higher than 8.0.[a]

7. When the bromophenol blue is close to the bottom of the gel, stop the electrophoresis and stain the gel in ethidium bromide (0.5 µg/ml in 25 mM Tris–HCl, pH 9.0) for 1 h with gentle rocking.[b] If background staining is high, the gel can be de-stained in water (two changes, each of 30 min).

8. Examine the gel on a UV transilluminator, and photograph it with a Polaroid camera.[c]

---

[a] It is important that the RNA samples are completely denatured before electrophoresis and kept denatured during electrophoresis, as the formation of secondary structures in the RNA molecules alters their electrophoretic mobility. Therefore, the quality of the reagents in the denaturing mixture and re-circulation of the gel running buffer during electrophoresis are crucial for the success of this method.

[b] Glyoxal reacts with ethidium bromide. The gel should therefore be prepared and run in the absence of the dye.

[c] In an un-degraded RNA preparation, the intensity of the 28S band should be at least twice that of the 18S band.

Protocol 11. Electrophoresis of RNA using a formaldehyde gel

Materials

All reagents should be RNase-free or DEPC-treated

- Formaldehyde: this is usually obtained as an approx. 37% solution in water. It oxidizes in air so bottles should be kept tightly closed. The pH of the concentrated solution should be greater than 4.0. If it is lower than 4.0, the solution should be discarded.

- Formamide: in general, most high-grade formamide can be used directly without any pre-treatment. However, if a yellow colour is observed, the formamide should be deionized by mixing it with a mixed bed resin as described in *Protocol 4*.

- 10 × electrophoresis buffer: 0.2 M MOPS, 75 mM sodium acetate, 20 mM EDTA, pH 7.0

- Loading buffer: 50% glycerol, 0.25% bromophenol blue in 1 × electrophoresis buffer

Method

1. Make a 1.6% (w/v) agarose solution in DEPC-treated water and cool it to approx. 60°C (see *Protocol 9*).

2. Add 1/7 vol. of 10 × electrophoresis buffer and 1/4 vol. of formaldehyde (37% solution) to the agarose solution and mix thoroughly. The final concentration of agarose is approx. 1.1%.

3. Dissolve each RNA sample (containing up to 20 μg of RNA) in 3.3 μl of water and add 1.5 μl of 10 × electrophoresis buffer, 7.5 μl of formamide and 2.7 μl of formaldehyde.

4. Denature the RNA samples at 65 °C for 15 min and chill them on ice for 5 min.

5. Add 2.5 μl of loading buffer to each sample[a], mix and spin by pulsing at 12 000 g in a microfuge.

6. Load the samples into the wells of the agarose gel[b] and run it at 1–3 V/cm in 1 × electrophoresis buffer. The gel running buffer should be changed after 1–2 h of electrophoresis.

7. After electrophoresis (when the bromophenol blue has migrated close to the bottom of the gel), stain the gel in ethidium bromide (0.5 μg/ml in water) for 1 h with gentle rocking. If the background staining is high, the gel can be destained in water (two changes of 30 min each).

8. Examine the gel on a UV transilluminator and photograph it with a Polaroid camera.

---

[a] Ethidium bromide (1 μl of a 1 mg/ml solution) can be added to the samples before electrophoresis so that the gel can be examined directly.
[b] Degradation of DNA fragments usually occurs during electrophoresis on a formaldehyde gel. Furthermore, RNA molecules migrate faster than DNA molecules of the same size under these conditions. Therefore, RNA (not DNA) molecular weight markers should be used with this gel system.

6.2 Quantitative analysis of nucleic acids

Two methods are widely used to determine the concentration of DNA and RNA in solution. In general, spectrophotometric measurement of UV absorption of the bases is simple and accurate. However, this method is relatively insensitive. If the total amount of nucleic acid is small, the concentration of DNA and RNA can be estimated from the intensity of UV-induced fluorescence emitted after ethidium bromide staining.

6.2.1 Spectrophotometric method

The concentration of nucleic acids is calculated from the optical density (OD) at 260 nm. 1 OD unit corresponds to approximately 50 μg/ml for double-stranded DNA and 40 μg/ml for single-stranded DNA and RNA. The ratio between the absorbances at 260 nm and 280 nm can be used to estimate sample purity. Pure preparations of DNA and RNA have ratios of 1.8 and 2.0 respectively. If the ratio is less than these values, the samples are contaminated with protein or phenol and the calculated concentration of nucleic acid is not then reliable. Under these circumstances, proteinase K digestion and phenol/chloroform extraction should be repeated (see *Protocols 2* and *3*).

6.2.2 Ethidium bromide staining

i. Gel electrophoresis method

One simple way to estimate the concentration of nucleic acids is to load different amounts of marker DNA or RNA on the gel when the samples are analysed qualitatively (see Section 6.1). The concentration of DNA or RNA can then be estimated by comparing the intensity of fluorescence of the samples with that of known concentrations of marker standards. However, it is important that DNA and RNA samples should only be compared with their counterpart markers because the same quantity of DNA and RNA emits a different intensity of UV-induced fluorescence.

ii. Spotting method

The concentration of DNA or RNA in solution can be estimated directly by staining the solution with ethidium bromide, measuring the resultant UV-induced fluorescence and comparing this with the fluorescence of known marker standards. As for the gel electrophoresis method, standards of the same nucleic acid type (that is, DNA or RNA) should be used. This procedure is described in *Protocol 12*.

Protocol 12. Estimation of DNA/RNA concentration by UV-induced fluorescence

Materials
- TE buffer: 10 mM Tris–HCl, 1 mM EDTA, pH 8.0
- Ethidium bromide solution: 1 μg/ml in TE buffer

Method

1. Make up serial dilutions of DNA or RNA samples and appropriate standard solutions in TE buffer and add an equal volume of ethidium bromide solution.
2. Stretch a sheet of plastic wrap (for example, Saran®) over the UV transilluminator.
3. Spot 5 μl of each sample and standard solution on to the plastic wrap.
4. Estimate the concentration of each sample by comparing the intensity of its fluorescence with those of the standards. A photograph may be taken with a Polaroid camera for easy estimation.

---

7. Storage of nucleic acids

- RNA should always be stored at −70°C in 70% ethanol.

- For short-term storage of DNA, 4°C is the best and simplest condition. For long-term storage, −70°C is more suitable. However, damage of DNA by free radicals can be greater when it is frozen.
- DNA should not be stored at −20°C because extensive single and double-strand breaks occur at this temperature. This is compounded by further strand breaks induced by freeze-thawing.

References

1. Blin, N. and Stafford, D. W. (1976). *Nucleic Acids Res.*, **3**, 2303.
2. Chan, V. T. W., Fleming, K. A., and McGee, J. O'D. (1988). *Anal. Biochem.*, **168,** 16.
3. Kupiec, J. J., Giron, M. L., Vilette, D., Jeltsch, J. M., and Emanoil-Ravier, R. (1987). *Anal. Biochem.*, **164,** 53.
4. Favorolo, J., Freisman, R., and Kamen, R. (1980). In *Methods in Enzymology* (ed. L. Grossman and K. Moldave), p. 718. Academic Press, New York.
5. Glisin, V., Crkvenjakov, R., and Byus, C. (1974). *Biochemistry*, **13**, 2633.
6. Chirgwin, J. M., Przybyla, A. E., McDonald, R. J., and Rutter, W. J. (1979). *Biochemistry*, **18,** 5294.
7. Aviv, H. and Leder, P. (1972). *Proc. Natl. Acad. Sci. USA*, **69,** 1408.
8. McMaster, G. K. and Carmichael, G. G. (1977). *Proc. Natl. Acad. Sci. USA*, **74,** 4835.
9. Lehrach, H., Diamond, D., Wozney, J. M., and Boedtker, H. (1977). *Biochemistry*, **16,** 4743.

2

Probe preparation and labelling

JULIA E. STICKLAND

1. Introduction

This chapter describes methods for the purification of recombinant bacterial
plasmids and the subsequent labelling of the cloned insert DNA, or RNA
copied from it, by a variety of enzyme-based techniques. Procedures for both
isotopic and non-isotopic labelling are given, together with a discussion of the
relative merits of each of the different labelling techniques.

2. Bacterial plasmids

2.1 Background

Bacterial plamids are double-stranded, closed circular DNA molecules that
range in size from 1 kb to more than 200 kb. They replicate and are inherited
independently of the bacterial chromosome, although they rely on enzymes
and proteins encoded by the host for their replication and transcription.

Under natural conditions, many plasmids are transmitted to new hosts by
bacterial conjugation. In the laboratory, plasmid DNA can be introduced into
bacterial by the artificial process of transformation (1). In this process, bacteria
are treated with mixtures of divalent cations to make them temporarily
permeable to small DNA molecules.

Selectable markers, encoded by the plasmid, are used to identify the
transformants. The most commonly used markers are genes that confer resistance to antibiotics such as ampicillin or tetracycline.

Over the years plasmid cloning vectors have been engineered to reduce
their size to a minimum, while expanding their capacity to accept fragments
of foreign DNA generated by cleavage with a wide range of restriction
enzymes. Practically all vectors now contain a closely arranged series of
unique (that is, not found elsewhere in the vector) synthetic cloning sites
termed 'polylinkers' or polycloning sites which consist of sequences recognized by restriction enzymes commonly used in cloning experiments. For
example, the polycloning site from the vector pUC19 consists of a tandem
array of cleavage sites for 14 restriction enzymes (2) (see *Figure 1*). By

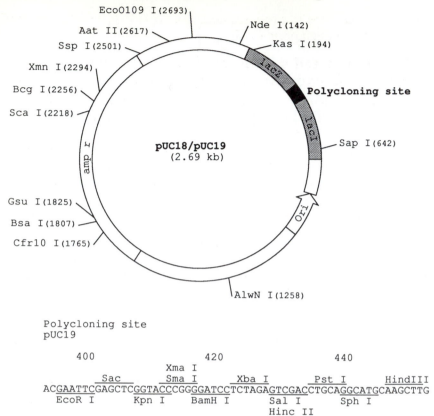

Figure 1. Restriction map of pUC18/pUC19 showing unique restriction enzyme sites. The polycloning sites for pUC19 are shown below the map.

incorporating the polycloning sites into a gene that codes for a biologically active fragment of the enzyme β-galactosidase, insertion of foreign DNA nearly always leads to the production of an inactive enzyme fragment. The recombinant clones lose their ability to hydrolyse a chromogenic substrate, 5-bromo-4-chloro-3-indolyl-β-D-galactosidase (X-gal), and can be distinguished from wild-type parental clones by a non-destructive histochemical test. Bacteria carrying recombinant plasmids give rise to white colonies whereas parental clones form blue colonies (a process termed α complementation) (3).

In addition to constructing plasmid vectors which allow visual identification of recombinant clones by histochemical tests, plasmids have been generated that allow: production of single-stranded templates for DNA sequencing [for example, pBluescript and pUC vectors (2)]; transcription of foreign DNA sequences *in vitro* [for example, pSP64/65 and their derivatives (4), and

26

pBluescript/pBluescribe (Stratagene, San Diego, USA)]; and expression of large amounts of foreign proteins (5).

2.2 Preparation and transformation of competent *Escherichia coli* cells by bacterial plasmids

Most methods for bacterial transformation are based on the observations of Mandel and Higa (6): bacteria treated with ice-cold solutions of $CaCl_2$ and then briefly heated, are induced into a transient state of 'competence' during which they are able to take up DNAs derived from a variety of sources.

Competent *E. coli* stocks may either be purchased from a commercial source (BRL, Gaithersburg, Maryland, USA) or prepared in the laboratory and used either fresh or from frozen stocks. Commercial products are very reliable, yielding transformation frequencies of $\geqslant 10^8$ colonies/μg of super-coiled plasmid DNA. They are, however, expensive.

A method of preparing competent bacteria that yields approximately 10^7 transformed colonies/μg of supercoiled plasmid DNA is detailed in *Protocol 1*. This method is ideal for use with such strains as, JM109, HB101, TG1, and TG2. Almost any *E. coli* strain can, however, be transformed at some frequency with circular plasmid DNA. These efficiencies of transformation are high enough to allow all routine cloning into plasmids, but if higher frequencies of transformation are required (for example, when constructing plasmid libraries of cDNAs synthesized from rare mRNA) alternative methods should be used.

A procedure developed by Hanahan (7) can yield competent cultures of *E. coli* strains DH1, DH5, and MM294 that can be transformed at frequencies of 7.5×10^8 colonies/μg of supercoiled plasmid DNA. Higher transformation efficiencies still, of $10^9–10^{10}$ transformants/μg of DNA, may be achieved, but only when using the method of electroporation. This method was developed originally to introduce DNA into eukaryotic cells but it has recently been used to transform *E. coli*. A special high-voltage mini-electrode and sample holder are required and these are available commercially from Bio-Rad together with a detailed protocol.

Several factors are critical in the protocol given below. Particular attention should be paid to:

(a) the density of cells at harvest, and

(b) the necessity for maintaining the cells at 0–4°C throughout processing.

Protocol 1. Preparation of competent *E. coli* cells, using calcium chloride

Materials
- 0.1 M $CaCl_2$
- Dimethylsulfoxide (DMSO)

Protocol 1. *Continued*

- LB medium: 10 g bacto-tryptone, 5 g bacto-yeast, 10 g NaCl. Make up to 1 litre with deionized water and adjust the pH to 7.0 with 5 M NaOH. Sterilize by autoclaving (15 lb/in^2) on a liquid cycle.

- For LB plates add 15 g of DIFCO agar per litre of LB medium. Sterilize by autoclaving as above.

- SOB medium: 20 g bacto-tryptone, 5 g bacto-yeast, 0.5 g NaCl. Make up to 950 ml with deionized water and add 10 ml of 250 mM HCl. Adjust the pH to 7.0 with 5 M NaOH and adjust the volume to 1 litre. Sterilize by autoclaving as above.

- SOC medium
  This is identical to SOB medium, except that it contains 20 mM glucose in addition. After the SOB medium has been autoclaved, allow it to cool and add 20 ml of 1 M filter-sterile glucose.

Methods

1. Pick a single bacterial colony from an LB agar plate, freshly grown for 16–20 h at 37°C, and inoculate 10 ml of LB medium in a 20-ml tube. Incubate at 37°C with vigorous aeration (200 r.p.m. in a rotary shaker) overnight.

2. Transfer 1 ml of the overnight incubation to 100 ml of LB medium in a 1-litre flask. Incubate at 37°C with vigorous aeration (200 r.p.m. in a rotary shaker) for approx. 2 h until the cell density is approx. 10^8/ml (that is, the OD_{600} is approx. 0.4).

3. Transfer the cells to a sterile, ice-cold 50-ml tube (Falcon 2070) and cool to 0°C by storing on ice for 10 min.

4. Recover the cells by centrifugation at 4300 g for 5 min at 4°C in a Sorvall GS3 rotor (or its equivalent).

5. Decant the medium from the cell pellets. Invert the tubes and allow to stand for 1 min to remove the last traces of medium.

6. Resuspend each pellet in 10 ml ice-cold 0.1 M CaCl$_2$ by gentle shaking and incubate on ice for 30 min.

7. Recover the cells by centrifugation at 4300 g for 5 min at 4°C in a GS3 rotor (or its equivalent).

8. Decant the supernatant from the cell pellets and stand the tubes in an inverted position for 1 min.

9. Resuspend each pellet in 2 ml ice-cold 0.1 M CaCl$_2$ for each 50 ml of original culture. Cells are best left for 1 h at 4°C before use and may be stored at 4°C for up to 48 h. (Transformation efficiency will drop after 24 h.) Cells may also be dispersed into aliquots and frozen at −70°C.[a]

10. Using a chilled sterile pipette tip, transfer 200 μl of each suspension to a sterile polypropylene tube (Falcon 2059 17 × 100 mm). Add supercoiled plasmid DNA to each tube (no more than 50 ng in a total volume of 10 μl or less), mix the contents by gentle swirling and store on ice for 30 min.

Controls:

- Competent bacteria that receive a known amount of a preparation of supercoiled plasmid DNA (1–20 ng).
- Competent bacteria that receive no plasmid DNA at all.

11. Transfer the tubes to a rack in a water bath at 42°C. Leave for exactly 90 sec. Do not shake the tubes.

12. Rapidly transfer the tubes back on to ice and leave for 2 min.

13. Add 800 μl of SOC medium to each tube. Incubate at 37°C with gentle aeration (100 r.p.m.) for 45 min to allow the bacteria to recover and express the antibiotic resistance marker encoded by the plasmid.

14. Transfer up to 200 μl of transformed competent cells on to 90 mm LB agar plates or SOB agar plates containing the appropriate antibiotic. Using a sterile, bent glass rod, gently spread the transformed cells over the surface of the agar plate.

15. Leave the plates at room temperature until the excess moisture has been absorbed.

16. Invert the plates and incubate at 37°C. Colonies should appear in 12–16 h.

---

a Storage of competent cells

(a) Add 70 μl of DMSO per 2 ml of resuspended cells.
Mix gently by swirling, and store the suspension on ice for 15 min.

(b) Add an additional 70 μl of DMSO to each suspension.
Mix gently by swirling, and place on ice.

(c) Immediately, dispense 50 μl aliquots of the suspensions into chilled sterile microfuge tubes and snap freeze the competent cells by immersing in liquid nitrogen.
Store the tubes at −70°C until required;

(d) When needed, thaw the cells by holding the tube in the palm of your hand. As the cells thaw, place on ice for at least 10 min.

2.3 Extraction and purification of plasmid DNA

2.3.1 Introduction

Many plasmid purification methods have been developed. The methods vary in the purity of the final product, the speed with which it may be obtained and the applicability of the purification procedures to large-scale cultures ('maxi-preps'), cultures of intermediate size ('midipreps'), and the simultaneous isolation of plasmid DNA from many small-scale cultures ('minipreps').

Plasmids are almost always purified from cultures grown in liquid medium containing the appropriate antibiotics that have been inoculated with a single bacterial colony picked from an agar plate. If the plasmid vectors (for example, the pUC series) replicate to a high copy number, they can be purified in large yield from cultures that have been grown to late log phase (approx. 18 h) in standard LB medium. However, vectors which do not replicate so freely (such as pBR322) need to be amplified selectively by incubating the bacterial culture in chloramphenicol (170 μg/ml) for several hours (8). Chloramphenicol inhibits host protein synthesis and, as a result, prevents replication of the bacterial chromosome. The replication of the plasmid, however, continues progressively for several hours giving considerably higher yields than obtained with unamplified cultures.

2.3.2 Basic procedures

Most plasmid isolation procedures make use of the covalently closed circular (CCC) form of plasmid DNAs and their small size in relation to the bacterial chromosome. Bacteria are recovered by centrifugation and lysed, normally by treatment with EDTA and lysozyme, which break down the cell wall. Lysis may also be achieved using organic solvents, heat, or non-ionic detergents (9). Cell debris and fragments of the bacterial chromosome are subsequently removed by centrifugation.

The alkaline extraction procedure devised by Birnboim and Doly (10) is one of the most useful for any scale of work, large or small. In this procedure, the bulk of contaminating chromosomal DNA, RNA, and protein are precipitated to yield plasmid preparations which are sufficiently pure for restriction endonuclease digestion and subsequent cloning procedures. Alkaline extraction can be followed by dye-buoyant density centrifugation (11), to further purify plasmids. This technique, which also depends upon the CCC form of plasmid molecules, gives high yields of very pure plasmid DNA which lacks chromosomal fragments.

i. Alkaline extraction

The principle of this technique is that alkaline conditions (approx. pH 12), which denature linear DNA molecules (that is, which separate the two strands of the host double helix), do not denature CCC plasmid molecules. Denatured chromosomal DNA precipitates when the cell extract is neutralized at high salt concentration, presumably because reassociation of long single-stranded DNA molecules occurs at multiple sites to form an insoluble mass. Cellular RNA also precipitates as does protein, since the detergent sodium dodecyl sulfate (SDS) is present in the reaction. A procedure for purifying small quantities of plasmid DNA from *E. coli* K12 strains is detailed in *Protocol 2*. A procedure for purifying larger quantities of more pure plasmids, but which includes a dye-buoyant density gradient, is listed in *Protocol 3*.

ii. *Dye-buoyant density centrifugation*

Significant differences in the buoyant density of plasmid and chromosomal DNA can be introduced by adding the dye ethidium bromide. When ethidium bromide binds to DNA, it causes the double helix to unwind by intercalation between the bases. This leads to an increase in the length of the linear chromosomal DNA molecules, thus decreasing their buoyant densities, and to the introduction of compensatory superhelical turns in CCC plasmid DNAs. Eventually, the intercalation of additional ethidium bromide molecules in CCC DNA is prevented because the number of superhelical turns becomes so great.

Linear fragments of chromosomal or plasmid DNA, which lack the constraints imposed on the CCC plasmids, bind more ethidium bromide and unwind further so their buoyant densities decrease more than that of CCC plasmids under the same conditions.

In ethidium-bromide-saturated caesium chloride density gradients, which are centrifuged to equilibrium, CCC plasmid DNA has a greater buoyant density than chromosomal DNA. As a result, CCC DNA forms a band below that formed by chromosomal DNA. Plasmid DNA is normally immediately visible as a red band without having to resort to long wavelength UV illumination. RNA forms a pellet at the bottom of the tube.

Following centrifugation, plasmids are removed from the tube by piercing the side with a needle attached to a syringe and collecting the plasmid fraction directly. For most purposes, the quantities of contaminating proteins and RNA remaining after dye-buoyant density centrifugation are insignificant. However, proteins can be removed by phenol/chloroform extraction (see *Protocol 10*) if they remain a problem, and RNA can be removed by a centrifugation or chromatography step if necessary (12). If easy access to an ultracentrifuge is not available, the large-scale purification protocol (*Protocol 3*) may be followed as far as step 8. Contaminating proteins should then be removed by following *Protocol 10* prior to ethanol precipitation.

Protocol 2. Small-scale isolation of plasmid DNA

Materials

- Solution I: 50 mM glucose, 25 mM Tris–HCl, pH 8.0, 10 mM EDTA. Store at 4°C.
- Solution II: 0.2 M NaOH, 1.0% (v/v) SDS. Use a freshly prepared solution. Store at room temperature. Best used within 1 week of preparation.
- Solution III: 60 ml 5 M potassium acetate, 11.5 ml glacial acetic acid, 28.5 ml water. The resulting solution is 3 M with respect to potassium and 5 M with respect to acetate.

Protocol 2. *Continued*

● Solution IV: 100 mM NaCl, 1 mM EDTA, 10 mM Tris–HCl, pH 7.5

Method

1. Transfer a single bacterial colony (the end-product of a transformation experiment from *Protocol 1*, or from a previously streaked bacterial glycerol stock) into 2 ml of LB medium[a] containing the appropriate antibiotic in a loosely capped 15-ml tube. Incubate the culture overnight at 37°C with vigorous shaking (200 r.p.m.) to give an OD_{600} of about 1.0.

2. Pour 1.5 ml of the culture into a microfuge tube. Centrifuge at 12 000 g for 1 min to harvest the cells. Store the remainder of the culture at 4°C.

3. Remove the medium by aspiration and resuspend the bacterial pellet in 100 µl of ice-cold solution I by vigorous vortexing. Store on ice for 15 min.

4. Add 200 µl of solution II. Mix the contents by inverting the closed tube rapidly five times. Do not vortex. Store the tube on ice for 5 min.

5. Add 150 µl of solution III. Close the tube and vortex gently for 10 sec. Store the tube on ice for 10 min.

6. Centrifuge at 12 000 g for 5 min at 4°C in a microfuge. Transfer the supernatant to a fresh tube.

7. Add 1 ml of ice-cold 100% ethanol to precipitate the double-stranded DNA. Mix by vortexing. Store at −20°C for 30 min.

8. Centrifuge at 12 000 g for 5 min at 4°C in a microfuge.

9. Carefully remove the supernatant and resuspend the pellet in 100 µl of solution IV.

10. Add 200 µl of ethanol. Vortex and place at −20°C for 10 min. Centrifuge at 12 000 g for 5 min at 4°C in a microfuge.

11. Discard the supernatant. Add 1 ml of 100% ethanol to the pellet. Centrifuge at 12 000 g for 2 min at 4°C in a microfuge.

12. Discard the supernatant and allow the pellet of nucleic acid to air-dry for 10 min. Redissolve the nucleic acids in 50 µl of TE buffer.[b] Store at −20°C.

[a] For composition of LB medium see *Protocol 1*.
[b] TE buffer is 10 mM Tris–HCl, pH 8.0, 1 mM EDTA

The typical yield of high-copy number plasmids such as pUC prepared by the method given in *Protocol 2* is 3–5 µg/ml of original bacterial culture. If the miniprep DNA isolated in this way is resistant to cleavage by restriction enzymes, extraction with phenol:chloroform may be necessary to remove impurities (see *Protocol 10*). The protocol can be scaled up appropriately to accommodate up to 10 ml of bacterial culture.

Protocol 3. Large-scale isolation of plasmid DNA

1. Inoculate 30 ml of LB-medium[a] (containing the appropriate antibiotic) with a 1/100 dilution of an overnight culture taken from a single bacterial colony containing the plasmid of interest.
   Grow the bacteria to late log phase (OD_{600} of approx. 0.6 usually achieved in 2–3 h), at 37°C with shaking (200 r.p.m.).

2. Inoculate 500 ml of LB-broth[a] (pre-warmed to 37°C) containing the appropriate antibiotic in a 2-litre flask, with 25 ml of the culture from Step 1. Incubate for 2.5 h at 37°C with vigorous shaking (200 r.p.m.). The OD_{600} of the resulting culture will be approx. 1.0.

   - Optional: For lower copy number plasmids (for example, pBR322) add 2.5 ml of a solution of chloramphenicol (34 mg/ml in ethanol) to give a final concentration in the culture of 170 µg/ml.

3. Incubate the culture for a further 12–16 h (conventionally overnight) at 37°C with vigorous shaking (200 r.p.m.).

4. Harvest the cells by centrifugation (10000 g for 10 min in the Sorvall GSA rotor or equivalent). Discard the supernatant. Stand the open centrifuge bottle in an inverted position to allow the supernatant to drain away.

5. Resuspend the bacterial cell pellet in 10 ml solution I[b] containing lysozyme (2 mg/ml) by shaking. Incubate for 10 min on ice.

6. Add 20 ml of solution II.[b] Close the top of the centrifuge bottle and mix the contents thoroughly by gently inverting/swirling the bottle several times. Store on ice for 5 min.

7. Add 20 ml of ice-cold solution III.[b] Close the top of the centrifuge bottle and mix the contents thoroughly by shaking several times. Store on ice for 10 to 20 min. A flocculent white precipitate of chromosomal DNA, RNA, and protein/membrane complexes should form.

8. Centrifuge the bacterial lysate at 10000 g for 10 min at 4°C in a GS3 rotor (or its equivalent).

9. Filter the supernatant through cheese-cloth into a 250-ml centrifuge bottle. Add 0.6 vol. isopropanol, mix well, and incubate at −20°C for 20 min to precipitate the nucleic acids.

10. Pellet the nucleic acids by centrifugation at 10000 g for 10 min at 4°C in a Sorvall GS3 rotor (or its equivalent).

Protocol 3. *Continued*

11. Decant the supernatant carefully and invert the bottle to allow the last drops of supernatant to drain away (approx. 30 min at room temperature).

12. Dissolve the nucleic acid pellet in 8.8 ml of TE buffer.[c]

13. To 8.8 ml of resuspended pellet, add 1.0 g of solid CsCl per millilitre of DNA solution. Mix gently until dissolved and add 0.4 ml of a solution of ethidium bromide (10 mg/ml in water).

14. Using a Pasteur pipette or disposable syringe, transfer the clear red solution to a Beckman quickseal (or equivalent) tube adding light paraffin oil to fill the tube. Seal the tube following the manufacturers instructions and centrifuge at 100 000 g for 40 h at 20°C in, for example, a Beckman Ti50 rotor. The equivalent fixed-angle or vertical rotors from other manufacturers could also be used (such as 250 000 g for 16 h in, for example, a Beckman 70Ti rotor). Tubes should be balanced to within 0.02 g.

15. To remove the plasmid DNA from the tube, insert a 19-gauge hypodermic needle into the top of the tube to allow air to enter. Then insert a 21-gauge needle (bevelled side up) attached to a 2-ml syringe into the lower red coloured band (if the yield of plasmid is low, illuminate with long wavelength UV light). Remove the plasmid DNA and transfer to a sterile tube.

16. Extract the ethidium bromide using butan-1-ol saturated with water: add an equal volume of saturated butan-1-ol to the plasmid DNA, mix, and discard the upper layer. Repeat until the butan-1-ol is no longer coloured.

17. Remove the CsCl from the DNA solution by diluting with an equal volume of water and precipitating the DNA with 2 vol. of ethanol for 15 min at 4°C. Recover the plasmid DNA by centrifugation at 10 000 g for 15 min at 4°C in a Sorvall SS34 rotor or equivalent. Dissolve the precipitated DNA in approx. 1 ml of TE buffer, measure the OD_{260} of the final DNA solution, and calculate the concentration of DNA (see Chapter 1). Store in aliquots at −20°C.

[a] For composition of these solutions see *Protocol 1*.
[b] For composition of these solutions see *Protocol 2*.
[c] TE buffer is 10 mM Tris–HCl pH 8.0, 1 mM EDTA.

3. Labelling of nucleic acids

3.1 Introduction

Techniques such as Southern and Northern blot analysis (see Chapter 3), dot blots (see Chapter 3), colony/plaque lifts, S1/RNase mapping (12), and *in situ* hybridization (see Volume I) (13), depend upon the ready availability of methods to introduce a label into cloned segments of DNA, or RNA copied from the cloned DNA. In each case, a combination of factors must be taken into account to determine which label and which labelling technique are applicable to the project in hand.

Broadly speaking, the choice of label is determined by the degree of sensitivity and resolution required, in combination with other factors such as probe stability, safety, ease of use and cost.

3.2 Radioactive versus non-radioactive labelling

Traditionally, DNA or RNA probes are prepared using radioactively [^{32}P], [^{35}S], [^{125}I], [^{3}H] modified nucleotides which, following hybridization to the target nucleic acid, can be detected by autoradiography. Most isotopes used for nucleic acid detection, however, have a short half-life ([^{32}P], 14.3 days and [^{35}S], 87.4 days), and therefore require frequent probe preparation. In addition, stringent safety precautions are required when working with isotopes and radioactive waste disposal is expensive. A search for effective non-radioactive markers for use in labelling reactions has recently provided at least two alternatives to working with isotopes: biotin–dNTP (bio–dNTP: Gibco BRL or Sigma) and digoxigenin–dUTP (dig–dUTP: Boehringer Mannheim). Some workers, however, find that the sensitivity, specificity, and reproducibility of these non-isotopic alternatives is not equal to that obtained with radioactivity, particularly for filter hybridization. This is, however, not the case for *in situ* hybridization (see Volume I).

Most non-radioactive methods involve enzymatic labelling using nucleoside triphosphates coupled covalently to a marker (usually biotin). Following hybridization, the marker can be detected using either non-immunological detection systems such as avidin and streptavidin (which have a very strong affinity for biotin), or by using antibodies to the reporter molecule which are coupled to either an enzyme or fluorescence markers. Incubation of the enzyme-coupled antibody with substrate results in visualization of the labelled hybrids (14).

It is also possible to label probes chemically, using biotin linked to highly reactive molecules. Biotinylated substrates [for example, photobiotin (15), biotin hydrazide (16), or a biotin ester (17)], can be made to react with one of the functional groups of nucleic acids (for example, amine group). The sensitivity and reproducibility of these methods, however, must be confirmed before they will successfully rival enzymatic labelling methods.

Unfortunately, even enzymatically labelled biotin probes may give rise to non-specific side-reactions, especially when used for *in situ* hybridization, as biotin itself is found in almost all natural materials. Furthermore, streptavidin or avidin used for detection of the hybrids tends to bind non-specifically to tissues and membranes (especially highly charged nylon membranes) resulting in increased background. However, there are many effective protocols for nucleic acid detection by *in situ* hybridization using biotinylated probes (see Volume I).

Labelling nucleic acids for use as hybridization probes with the cardenolide digoxigenin appears to be more specific, reproducible, and comparable in sensitivity to radiolabelled probes (18). Digoxigenin is only present in Digitalis plants and therefore reduces non-specific staining due to endogenous products.

For the purposes of this chapter, fast and reliable protocols for enzymatic labelling of nucleic acids with radioactive isotopes are given. Alterations for use with the non-radioactive reporter molecules biotin or digoxigenin are given at the end of each protocol. DNA labelling kits are available from companies such as Pharmacia and Amersham for use with radioisotopes, and from BRL and Boehringer Mannheim for non-radioactive labelling.

3.3 Basic procedures

All of the methods are straightforward enzyme catalysed reactions. However, some general points may help towards obtaining good reproducible labelling.

Reagents should be added in the order described in the protocol. Label is added penultimately to reduce handling of radioactive solution and is followed by addition of enzyme. Thorough mixing of all reagents before addition of enzyme is important (brief vortexing followed by centrifugation at $12\,000\,g$ for 2 sec in a microfuge). Enzyme is mixed in by gently pipetting two or three times. Poor labelling is normally due to impure DNA (or RNA). Phenol extraction followed by ethanol precipitation (see *Protocol 10*) is usually of benefit if this is the case. In addition, it is often possible to improve labelling efficiency by the use of more enzyme or by extending the reaction time.

When establishing a labelling method or using a new preparation of substrate nucleic acid, it is valuable to determine the efficiency of labelling. For radioactively labelled probes, *Protocol 4* gives a procedure for the most commonly used of these techniques: trichloroacetic acid (TCA) precipitation. Duplicates are usually taken at each time point, or at the end of the reaction, and the background of TCA insoluble counts is measured before enzyme is added. Once a method of labelling has been established, however, these steps can be omitted. For non-radioactive probes, the percentage incorporation of label and the average size of labelled probe fragments is determined by comparison to known labelled standards using Southern blot analysis (see Chapter 3 for methods). If required, unincorporated label can be removed as described in *Protocol 5*. In our experience, less background is observed on filter hybridizations and in *in situ* hybridizations if this procedure is carried out.

Protocol 4. TCA precipitation to estimate percentage incorporation of nucleotide triphosphate into DNA

Materials
- Whatman 540 filter paper
- 5% trichloroacetic acid/1% sodium pyrophosphate (TCA solution) **Care:** corrosive acid

Method
1. Spot 0.5 μl of the labelling reaction on to a small piece of Whatman 540 paper (labelled with a pencil).
2. For probes labelled with ^{32}P measure the radioactivity on the filter by Cerenkov counting (no scintillant). This gives the total counts in the sample. (If ^{35}S or ^{3}H incorporation is being assayed, the filters should be dried and counted with scintillant. Duplicate filters should be used for the TCA wash.)
3. Wash the same filter in 20 ml of TCA solution by shaking in a screw-capped disposable tube. Pour off the TCA solution and replace with fresh. Repeat the washes at least five times until no more radioactivity is washed off the filter.
4. Blot the filter dry and again measure the radioactivity by Cerenkov counting, to obtain the acid insoluble counts (the incorporated label).
5. Calculate the percentage incorporation as follows:
   Percentage incorporation

$$= \frac{\text{Acid visible counts (precipitated)}}{\text{Total counts}} \times 100$$

   Specific activity

$$= \frac{\text{Total incorporated counts}}{\text{Mass of hybridizable nucleic acid}}$$

Protocol 5. Removal of unincorporated nucleotides

Two methods are described for the removal of unincorporated nucleotides.

A. Ethanol precipitation

Materials
- 200 mM EDTA, pH 8.0
- 2.5 M sodium acetate, pH 5.0

Protocol 5. *Continued*

- Carrier (DNA or RNA or glycogen at a concentration of 10 mg/ml)
- Ice-cold ethanol
- TE buffer: 10 mM Tris–HCl, pH 8.0, 1 mM EDTA

Method

The method given is for a labelling reaction of 50 μl. For other volumes the reagents should be scaled up or down accordingly.

1. To a 50 μl labelling reaction, add the following:

   - 0.2 M EDTA, pH 8.0   5 μl
   - 2.5 M Na acetate, pH 5.0   5 μl
   - Carrier (20 μg)   2 μl
   - Ice-cold ethanol   200 μl

   Mix and chill to −70°C on dry ice or place at −20°C overnight.

2. Centrifuge at 12 000 g for 15 min in a microfuge to pellet the nucleic acid. Carefully remove the supernatant (if radioactive, transfer to a waste bottle for disposal).

3. Wash the pellet in ice-cold absolute ethanol, or 70% ethanol, and re-centrifuge as above. Remove the supernatant and allow the resulting pellet to air-dry.

4. Resuspend the labelled DNA pellet in the required volume of TE for use as a probe.

B. Sephadex spin columns

Materials

- 1-ml disposable syringe
- Sephadex G50-300 equilibrated in column wash buffer
- Carrier (DNA/RNA/glycogen 10 mg/ml)
- Column wash buffer: 10 mM EDTA, 150 mM NaCl, 0.1% SDS, 50 mM Tris–HCl, pH 7.5

Method

1. Plug the bottom of a 1-ml disposable syringe with a small amount of sterile cotton wool.

2. Prepare a column by filling the syringe to the top with Sephadex G50-300, taking care not to introduce air-bubbles.

3. Insert the syringe into a 15-ml tube, add 5 μl of carrier and wash through with 4 × 250 μl aliquots of column wash buffer.

4. Centrifuge at 1600 g for 4 min. Pour away the eluate from the bottom of the centrifuge tube.

5. Insert a 1-ml screw cap tube into the bottom of the centrifuge tube.

6. Apply the sample to the column and centrifuge at 1600 g for 4 min.

7. Collect the 1-ml tube containing the eluted probe (the unincorporated nucleotides remain in the syringe).

4. Methods for labelling double-stranded DNA

4.1 Nick-translation

The nick-translation reaction (19) involves the simultaneous action of two enzymes, pancreatic deoxyribonuclease I (DNase I) and *E. coli* DNA polymerase I (DNA pol I). DNase I introduces nicks at random points in both strands of a DNA duplex, producing a free 3'-hydroxyl and a free 5' phosphate group at each nick. The 5'–3' exonuclease activity of DNA polymerase I then progressively removes nucleotides from the 5' side of the nick. The simultaneous 5'–3' polymerase activity of DNA polymerase I adding nucleotides to the free 3'-hydroxyl ends results in movement of the nick (nick-translation) along the DNA. The process is illustrated diagrammatically in *Figure 2*. By replacing one or more of the pre-existing nucleotides with radioactive nucleotides in the reaction, the DNA can be labelled to high specific activity. Most of the parameters of the nick-translation reaction may be manipulated in order to alter probe size, specific activity, or yield.

The procedure in *Protocol 6* is appropriate for many situations. The example given in the protocol uses ^{32}P-labelled dNTPs but ^{3}H- and ^{35}S-radiolabelled nucleotides are also incorporated efficiently, as are the non-radioactive markers biotin and digoxigenin (see Volume 1).

Efficiencies of incorporation of over 60% should be achievable with typical specific activities of probe (d.p.m./µg) of 5×10^8.

Protocol 6. Labelling DNA by nick-translation

A. Radioactive

Materials

- 10 × nick-translation buffer (10 × NT buffer): 500 mM Tris–HCl, pH 7.6, 50 mM MgCl$_2$, 10 mM 2-mercaptoethanol
- TE buffer: 10 mM Tris–HCl, pH 8.0, 1 mM EDTA
- Stop buffer: 300 mM EDTA, pH 8.0

Protocol 6. *Continued*

- 10 × dNTP mixture: 200 μm each of dATP, dGTP, and dTTP in TE buffer, pH 8.0

- DNase I, 100–500 pg/μl

- DNA polymerase I, 2 units/μl

- [α-^{32}P]dCTP (specific activity >3000 Ci/mmol; 110 TBq/mmol, 10 μCi/μl)

Method

1. Add the following on ice to a 0.5-ml microcentrifuge tube:

    - DNA to be labelled   50–500 ng
    - 10 × NT buffer   5 μl
    - 10 × dNTP mixture   5 μl
    - Water to give a **final** total volume of   50 μl
    - [α-^{32}P]dCTP   5 μl
    - DNase I   2 μl (0.4–1.0 ng/μg DNA)
    - DNA polymerase I   2 μl (4 units)

    Mix by gently vortexing. Centrifuge at 12 000 g briefly (2 sec) to collect contents at the bottom of the tube.

2. Place the tube in a constant temperature water bath at 15°C for approximately 2 h.

3. Terminate the reaction by the addition of 5 μl of stop buffer and remove the unincorporated labelled nucleotides by ethanol precipitation or chromatography on Sephadex (see *Protocol 5*).

4. Denature the probe by heating to 95–100°C for 5 min and immediately place on ice for 5 min before use in a hybridization reaction. Probes may be stored at −20°C for several days before use.

B. Non-radioactive labelling

Materials
For biotin:

- 10 × dNTP mixture: 200 μm each dGTP, dCTP, and dTTP in TE buffer, pH 8.0

- Biotin-7-dATP (bio-7-dATP), 0.4 mM

For digoxigenin:

- 10 × dNTP mixture: 200 μm each dATP, dGTP, dCTP, and 130 μm dTTP in TE buffer, pH 8.0

- Dig-11-dUTP, 1 mM

Method

1. Add the following on ice to a 0.5 ml microcentrifuge tube:

- DNA to be labelled   1 µg
- 10 × NT buffer   5 µl
- 10 × dNTPs   5 µl
- 0.4 mM bio-7-dATP or   2.5 µl
  1.0 mM dig-11-dUTP   1 µl
- Water to a **final** volume of 50 µl
- DNase I   2 µl (0.4–1.0 ng/µg DNA)
- DNA polymerase I   2 µl (4 units)

Mix by gently vortexing. Centrifuge at 12 000 g briefly (2 sec) to collect contents at the bottom of the tube. Follow as for radioactive probes, step 2 onwards. Store small aliquot of probes at −70°C for several months, but keep working stock at 4°C.

The nick-translation reaction is not appropriate for single-stranded DNA, but it is equally efficient with both linear and circular double-stranded molecules. The concentration of DNase I determines the frequency of nicks and hence the final single-strand length of the labelled product. The procedure given in the above protocol results in single-strand lengths of 200–500 nucleotides, an optimum length for filter hybridization reactions.

There is a linear relationship between radioactive label incorporation and time over several hours.

4.2 Random primer labelling by primer extension

In common with nick-translation, primer extension methods utilize the ability of DNA polymerases to synthesize a new DNA strand complementary to a template strand starting from a free 3′-hydroxyl group. In random primer labelling, a free 3′-hydroxyl is provided by a short oligonucleotide primer annealed to the template (see *Figure 3*). If the oligonucleotides are heterogeneous in sequence, they will form hybrids at many positions so that every nucleotide of the template will be copied at equal frequency into the product. Hexanucleotides of random sequence, either derived from DNase I digestion of calf thymus DNA (Pharmacia, Boehringer Mannheim) or produced by oligonucleotide synthesis (International Biotechnologies) may be used to prepare labelled copies of both DNA and RNA.

The type of DNA dependent polymerase used depends on the nature of the template, but it is essential to use polymerase lacking 5′ to 3′ exonuclease activity, otherwise degradation of the primer will occur. RNA dependent

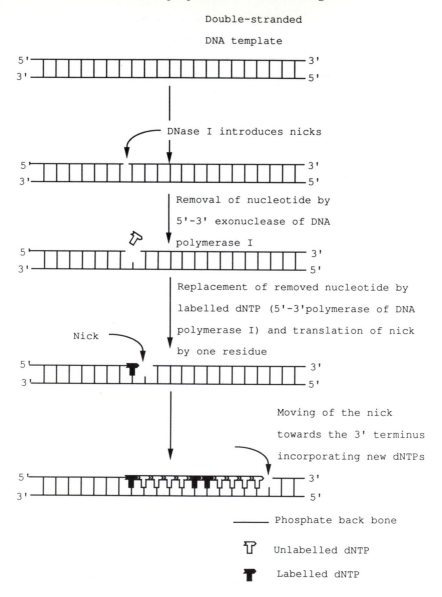

Figure 2. Nick-translation of DNA.

DNA polymerase (reverse transcriptase) is used to copy single-stranded RNA templates, whereas the Klenow fragment of *E. coli* DNA pol I is used when the template is single-stranded DNA.

Feinberg and Vogelstein first described the random primer labelling of DNA fragments using Klenow polymerase (20). The reaction is illustrated

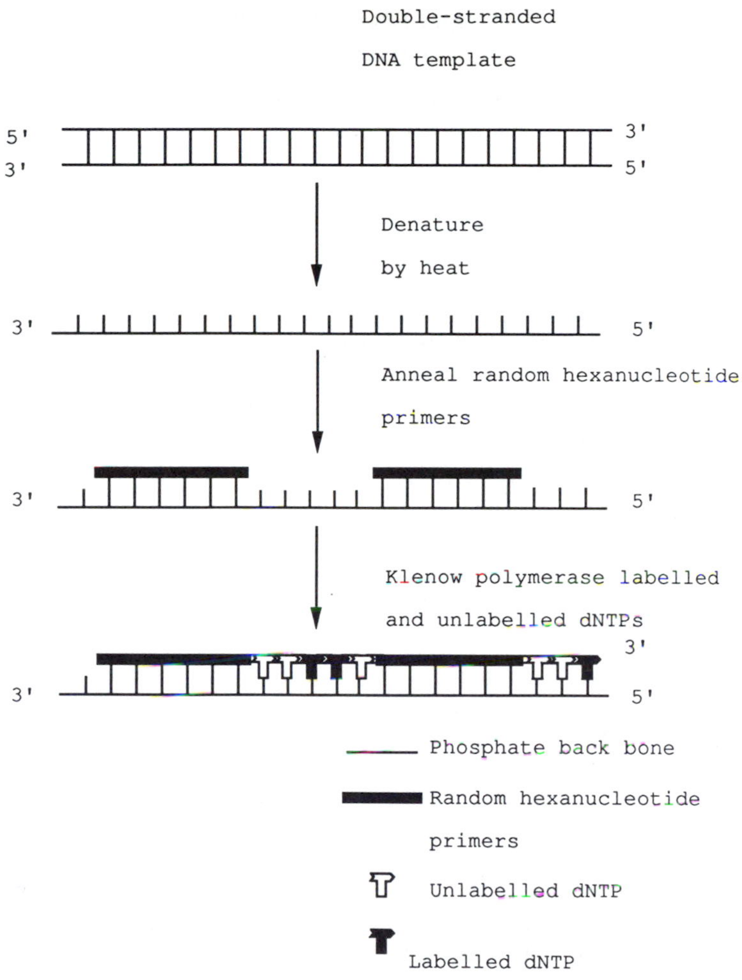

Double-stranded

DNA template

Figure 3. Random primer labelling of DNA.

diagrammatically in *Figure 3*. The procedure for labelling between 25–250 ng of single-stranded DNA is given in *Protocol 7*. The single-stranded DNA template may be derived either from cloning into a suitable single-stranded bacteriophage (for example, M13) or more frequently from heat denaturation of a double-stranded molecule. It is preferable to use linear molecules to avoid rapid renaturation of the complementary DNA circles. Relatively impure DNA may be labelled by random priming; for example, specific DNA fragments purified by agarose gel electrophoresis, a procedure frequently used to separate cloned insert from a vector sequence, which may show some cross-hybridization with the target DNA. Plasmid DNAs prepared by either the 'maxi' or the 'mini' prep methods outlined in this chapter, may be restricted with the appropriate enzymes, fractionated on an agarose gel and the inserts isolated and purified using one of a variety of methods (12). Alternatively, Feinberg and Vogelstein have also demonstrated that it is possible to label DNA fragments in the presence of low gelling-temperature agarose without prior purification (21). A protocol for preparation of DNA fragments in agarose is given in *Protocol 8*.

Protocol 7. Preparation of probes by random primer labelling

A. Radioactive

Materials

- $10 \times$ labelling buffer: 500 mM Tris–HCl, pH 8.0, 100 mM $MgCl_2$, 1 mM dithiothreitol (DTT), 2 mg/ml BSA
- Hexamers: 62.5 A_{260} units/ml
- $10 \times$ dNTP mixture: 200 µm each of dTTP, dATP, and dGTP in TE buffer, pH 8.0
- TE buffer: 10 mM Tris–HCl, pH 8.0, 1 mM EDTA
- $[\alpha\text{-}^{32}P]$ dCTP (specific activity >3000 Ci/mmol; 110 TBq/mmol, 10 µCi/µl)
- DNA polymerase I (Klenow fragment) 2 units/µl
- Stop buffer: 300 mM EDTA, pH 8.0

Method

1. Dissolve the DNA to be labelled in distilled water or TE buffer. Denature double-stranded DNA by heating to 95–100°C for 5 min and then chill on ice.

2. Add the following to a small microfuge tube on ice:
   - Denatured DNA   25–250 ng
   - $10 \times$ dNTP mixture   2 µl
   - $10 \times$ labelling buffer   2 µl
   - $[\alpha\text{-}^{32}P]$dCTP   5 µl (50 µCi, 2.0 MBq)
   - Water to a **final** volume of   20 µl
   - Klenow fragment   1 µl (2 units)

   Mix gently by slowly pipetting up and down. Centrifuge briefly (2 sec) in a microfuge to collect contents at the bottom of the tube.

3. Incubate at between 20°C and 37°C for 2–16 h.

4. Terminate the reaction with $2\,\mu l$ of stop buffer and remove unincorporated labelled nucleotides by ethanol precipitation or chromatography on Sephadex (see *Protocol 5*).

5. Denature the labelled DNA by heating to 95–100°C for 5 min and immediately place on ice for 5 min before use in a hybridization reaction.

B. Non-radioactive

Materials

As above, with the following changes:

For biotin:

- $10 \times$ bio–dNTP mixture: 1 mM each dATP, dCTP, dGTP, and 0.65 mM dTTP/0.35 mM bio-7-dUTP at pH 6.5

For digoxigenin:

- $10 \times$ dig–dNTP mixture: 1 mM each dATP, dCTP, dGTP, and 0.65 mM dTTP/0.35 mM dig-11-dUTP at pH 6

Method

1. Add the following to a microcentrifuge tube on ice:

• Freshly denatured DNA	10 ng–3 µg
• $10 \times$ labelling buffer	2 µl
• $10 \times$ dig–dNTP mix or $10 \times$ bio–dNTP mix	2 µl
• Water to a **final** volume of	20 µl
• Klenow fragment	1 µl (2 units)

Mix gently by slowly pipetting up and down. Centrifuge briefly (2 sec) in a microfuge to collect contents at the bottom of the tube.

2. Follow as for radioactive probes, step 3 onwards.

Protocol 8. Labelling of DNA fractionated by electrophoresis in low-melting-point agarose

1. After electrophoresis (in Tris–acetate buffer) in a suitable low-melting-point agarose gel (see Chapter 1 for agarose gel electrophoresis) containing 0.5 µg/ml ethidium bromide (**Caution**: EtBr is a powerful mutagen), excise the desired band (at least 250 ng or more of DNA).

2. Transfer the band, with minimum excess agarose, to a pre-weighed 1.5-ml microfuge tube.

Protocol 8. *Continued*

3. Add 3 ml of water for every gram of agarose gel and place the tube in a boiling water bath for 7 min to melt the gel and denature the DNA.

4. Transfer the tube to a water bath at 37°C for at least 10 min.

5. To a fresh tube, add the volume of DNA/agarose solution that contains 25 ng of DNA to the standard random primer labelling reaction. If this volume exceeds 10 μl then scale up the reaction to 50 μl with additional water. The reaction constituents may appear to gel during incubation, but the extension reaction will still proceed if this happens.

The concentration of radiolabel given in *Protocol 7* is appropriate for a wide variety of applications and a specific activity of 1.0×10^9 d.p.m./μg should be achieved using 25 ng of template, corresponding to 65–75% incorporation. Other radiolabels may be used during random priming but specific activities are significantly lower. For example, when using $[\alpha\text{-}^{35}\text{S}]\text{dATP}$ the specific activity for labelling 25 ng of template is approx. 7.7×10^8 d.p.m./μg.

Radiolabelled probes prepared by random priming are usually 400–600 nucleotides in length as determined by electrophoresis through alkaline agarose gels. With probes labelled to high specific activity with ^{32}P, radiolysis occurs very rapidly and storage of probes, even for 24 h, will lead to a significant reduction in mean probe length.

DNA fragments isolated from low-melting-point agarose are also labelled efficiently using biotin or digoxigenin. The labelling reaction results in digoxigenin/biotin incorporation every 20–25 nucleotides in the newly synthesized DNA, which gives the highest sensitivity in the detection reaction. The size of the labelled fragments is, depending on the template DNA, in the range of 200–2000 bp.

4.3 Probe generation by the polymerase chain reaction

The polymerase chain reaction (PCR) is an *in vitro* method for the production of large amounts of a specific DNA fragment from small amounts of a complex, often mixed template (22) (see Chapters 4, 6, and 7). By including a radioactive triphosphate during the PCR reaction, the amplified DNA, which can be anything from approx. 180 bp to 2.0 kb in size, can be labelled to a high specific activity of approx. 5×10^9 c.p.m./μg. This is greater than can be achieved by nick-translation or random primer labelling. Nick-translation normally generates a probe specific activity of 0.5×10^9 c.p.m./μg, but its efficiency of labelling declines for fragments less than 1 kb. Random primer labelling produces a specific activity of 2×10^9 c.p.m./μg and requires less template than nick-translation, but its labelling efficiency also declines for segments outside of the 0.5–2.0 kb size range. Probe generation by PCR is therefore particularly useful for labelling subnanogram amounts of template

DNA, less than 500 bp in length, to a very high specific activity. If required, lower specific activity probes can be generated using PCR by replacing a proportion of the labelled nucleotide with an equivalent amount of 'cold', or unlabelled nucleotide. Equivalent molarities of all four nucleotides must be present in the PCR reaction mixture to ensure the synthesis of full-length products.

Generally, templates for PCR labelling have already been subcloned into a vector. Primers can therefore be synthesized complementary to the regions just flanking the insertion sites of the vector, thereby allowing many segments to be labelled with a limited set of primers. More recently, however, PCR has also been used to generate labelled probes directly from genomic DNA (24).

It is possible to add DNA sequences, non-complementary to the template, to the 5' end of the oligonucleotide primers. These sequences then become incorporated into the double-stranded PCR product at the ends of the amplified target sequence. If these sequences represent a bacteriophage promoter, the amplified fragment can be transcribed using a bacteriophage encoded DNA-dependent RNA polymerase (see Section 5). Thus, a labelled RNA transcript can be produced without having to resort to subcloning the DNA segment next to a 'phage promoter (24).

4.3.1 Thermal cycling

PCR is performed by incubating the samples at three different temperatures corresponding to the three steps in a cycle of amplification–denaturation (90–95 °C), annealing (40–60 °C) and extension (72 °C). The equipment used for thermal cycling may be as simple as water baths set at different temperatures for manual reactions or as sophisticated as a microprocessor-controlled rapid-temperature-changing heat block for fully automated amplifications (available from, for example, Perkin-Elmer Cetus Instruments, Techne, and Koch-light).

The method for PCR probe labelling is detailed in *Protocol 9*. Insufficient heating during the denaturation step is a common cause of failure in the PCR reaction. To provide a margin of assurance, the reactants should be heated to about 93 °C, and then cooled rapidly to the annealing temperature. An overlay of about 50 µl of mineral oil is used to prevent loss of the sample by evaporation.

The annealing temperature depends on the length and G–C content of the primers, and a rough idea of the temperature can be derived using the formula:

$$4°C \times (G + C) + 2°C \times (A + T) - 3.$$

In practice, however, the temperature has to be determined empirically for each set of oligonucleotide primers (typically ~20 bp in length). Since the primers are present in a large molar excess in the reaction mix, hybridization occurs almost instantaneously, so only a short incubation is required at the annealing temperature.

Primer extension at 72°C is very close to the optimum temperature for *Taq* DNA polymerase activity. The incubation time at 72°C depends on the length of the DNA segment being amplified. For every 1000 bp of target, approximately 1 min should be allowed. If the target sequence is 150 bp or less the primer extension step can be eliminated altogether: the sample will be within the 70–75°C range, during its transition from annealing to denaturation temperature, for the few seconds required to completely extend the annealed primers.

Efficient synthesis of digoxigenin-labelled DNA by the PCR reaction has been reported, together with a simple procedure for purification of the non-isotopically labelled fragments using a low-melting-point agarose gel (25).

4.3.2 Precautions

Since the polymerase chain reaction is capable of amplifying a single molecule of DNA, precautions should be taken to guard against contamination of the reaction mixture with trace amounts of DNAs that could serve as templates (see Chapters 4 and 6 for details).

Protocol 9. Probe labelling by PCR

A. Radioactive

Materials
- 20 × reaction mixture (20 × RM): 1 M KCl, 200 mM Tris–HCl, pH 8.3, 30 mM MgCl$_2$, 0.2% (w/v) gelatin, 4 mM each dATP, dGTP, and dTTP
- Template DNA
- Primer mix: 20 µM of each of the 5' and 3' primers
- [α-^{32}P]dCTP (specific activity >3000 Ci/mmol; 110 TBq/mmol, 10 µCi/µl)
- *Taq* DNA polymerase 2 units/µl
- Mineral oil
- Stop buffer 300 mM EDTA, pH 8.0

Method
1. Set up the following reaction in a 500 µl microfuge tube at room temperature, in the order given:
    - 20 × RM   1 µl
    - Water   1 µl
    - Primer mix   2 µl
    - [α-^{32}P]dCTP   15 µl (150 µCi)
    - Template DNA   1 µl (2 ng)

2. Incubate at 93°C for 10 min, and then add:

Taq DNA polymerase   0.5 µl (1 unit)

3. Carry out 30–35 cycles of PCR at the specific annealing, extension and denaturation temperatures previously empirically determined for the primers being used.

An example of a PCR profile is given below:

annealing: 1 min at 50°C
extension: 2 min at 72°C
denaturation: 1 min at 93°C

4. Incubate at 72°C for 10 min for a final extension reaction.

5. Terminate the reaction with 2 µl of stop buffer and remove unincorporated labelled nucleotides by ethanol precipitation or chromatography on Sephadex (see *Protocol 5*).

6. Denature the labelled DNA by heating to 95–100°C for 5 min and immediately place on ice for 5 min before use in a hybridization reaction.

B. Non-radioactive

Materials

As above, with the following changes:

For biotin:

- 20 × reaction mixture (20 × RM): 1 M KCl, 200 mM Tris–HCl, pH 8.3, 30 mM $MgCl_2$, 0.2% (w/v) gelatin, 4 mM each dATP, dCTP, and dGTP; 2.6 mM dTTP, 1.4 mM bio-11-dUTP

For digoxigenin:

- 20 × reaction mixture (20 × RM): 1 M KCl, 200 mM Tris–HCl, pH 8.3, 30 mM $MgCl_2$, 0.2% (w/v) gelatin, 4 mM each dATP, dCTP, and dGTP; 2.6 mM dTTP, 1.4 mM dig-11-dUTP)

Method

1. Set up the following reaction in a 500 µl microfuge tube at room temperature, in the order given:

- 20 × RM   1 µl
  (dig-11-dUTP or bio-11-dUTP)
- Primer mix   2 µl
- Template DNA   1 µl (2 ng)
- Water   16 µl

Follow as for radioactive probes, step 2 onwards. Efficiency of label incorporation is detected as described in Section 3.3.

5. Synthesis of RNA probes by *in vitro* transcription

5.1 Bacteriophage encoded DNA-dependent RNA polymerases

RNA polymerases catalyse the polymerization of ribonucleoside triphosphates (rNTPs) into RNA using a DNA template. The RNA polymerases encoded by *E. coli* bacteriophages (for example, T7, T3, and SP6), are capable only of transcribing from particular promoters contained on the 'phage DNA (26). They do not recognize bacterial or plasmid promoters or eukaryotic promoters in cloned DNA sequences. Consequently, 'phage encoded RNA polymerases can be used to synthesize specific single-stranded RNA molecules *in vitro* (27).

5.2 Choice of vector

The specific transcription of a template DNA sequence requires cloning of the DNA sequence downstream of an appropriate 'phage promoter in a suitable vector (see *Figure 4*). The requirement of a specific cloning step has led to the development of a number of vectors in which a bacteriophage promoter is present upstream of a multiple cloning site (for example, pSP70 series (4), pBluescribe and pBluescript [Stratagene], and the pGEM series [Promega, Madison]). Choice of vector is largely a matter of personal preference. If transcripts of both strands of the template are required, however (that is, sense and antisense), it is more efficient to choose a plasmid carrying two different bacteriophage promoters than to insert the template in two orientations into a plasmid carrying a single promoter. This is of great help in the production of probes for the technique of *in situ* hybridization, because sense RNA can be used as a negative control in the localization of specific mRNA species (see Volume I).

It is important to consider the position of restriction sites within the template DNA and downstream from it. The 5' terminus of the transcript is fixed by the bacteriophage promoter, but the 3' terminus is determined by the downstream site chosen for cleavage with a restriction enzyme. A restriction enzyme that generates blunt or protruding 5' termini should be chosen if possible, since RNA polymerases can 'back-transcribe' from 3' overhanging tails, resulting in the synthesis of transcripts complementary to the intended probe (28).

5.3 Preparation of template DNA

Great care should be taken when working with RNA samples to avoid contamination of solutions with RNases. Gloves should be worn for all

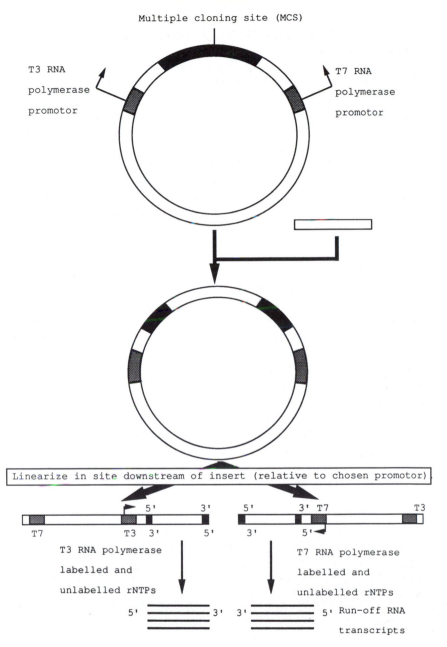

Figure 4. Preparation of RNA probes using a plasmid containing T7 and T3 RNA polymerase promoters.

manipulations and solutions treated with diethyl pyrocarbonate (DEPC). Add DEPC to 0.1% (v/v), leave for 1 h and then autoclave to destroy the DEPC. Tris buffers must not be treated with DEPC but should be made up with DEPC-treated water and autoclaved. Plasmid DNAs for use in transcription reactions should ideally be prepared by density gradient centrifugation (*Protocol 3*). However, crude 'minipreps' do work.

It is important that template DNA be cleaved to completion since supercoiled DNAs are efficient templates for all 'phage-encoded RNA polymerases and generate very long transcripts.

Following restriction enzyme digestion the DNA may be purified by phenol extraction, chloroform extraction, and finally ethanol precipitation (see *Protocol 10*).

Protocol 10. General method for extracting proteins from nucleic acids

1. *Phenol extraction*

   Add an equal volume of phenol saturated with TE[a] buffer (see Chapter 1) to the DNA or RNA solution. Mix by inversion until homogeneous. Separate the phases by centrifugation in a microcentrifuge at 12 000 g for 10 min. Carefully remove the upper (aqueous) phase with a pipette and place it in a clean tube.

2. *Phenol: chloroform extraction*

   Add an equal volume of phenol: chloroform: isoamylalcohol (25:24:1 v/v) to the aqueous phase from step 1. In practice, chloroform: isoamylalcohol (24:1 v/v) is made as a stock reagent and added to buffer-saturated phenol in a ratio of 1:1. Mix and separate the phases as described in step 1.

3. *Chloroform extraction*

   Extract the aqueous phase from step 2 with an equal volume of chloroform: isoamylalcohol (24:1). Separate the phases as described in step 1.

4. *Ethanol precipitation*

   Add 1/10 vol. 3 M sodium acetate pH 5.4 and 2 vol. ethanol to the aqueous phase from step 3. Mix and chill to −70°C or on dry ice.

   Pellet the nucleic acid by centrifugation in a microfuge for 10 min at 12 000 g. Carefully pour off the supernatant. Wash the pellet at room temperature with 70% ethanol, and re-centrifuge as above. Remove the excess ethanol and allow the pellet to dry in air. Resuspend in the required volume of TE[a] buffer.

[a] TE is 10 mM Tris–HCl, pH 7.5, 1 mM EDTA.

5.4 Transcription reaction

A method for *in vitro* transcription of high specific-activity, full-length RNA probes using [α-^{32}P]UTP is given in *Protocol 11* (27), as is a procedure for non-radioactive labelling with biotin or digoxigenin.

Protocol 11. Synthesis of RNA probes

A. Radioactive

Materials

- 10 × transcription buffer (10 × TB): 400 mM Tris–HCl, pH 7.5, 60 mM MgCl$_2$, 10 mM spermidine
- 10 × rNTPs: 5 mM each of rATP, rCTP, and rGTP, pH 7.5, 250 μM rUTP, pH 7.5
- [α-^{32}P]UTP (specific activity 400–3000 Ci/mmol)
- 100 mM DTT
- Placental ribonuclease inhibitor (RNasin), 40 units/μl
- SP6, T7, or T3 polymerase: 10–20 units/μl
- Template DNA linearized downstream of a bacteriophage promoter (0.1–1.0 μg)

Method

1. Set up the following reaction in a microfuge tube at room temperature in the order given:
   - 10 × TB   5 μl
   - 100 mM DTT   5 μl
   - RNasin   1.5 μl (approx. 80 units)
   - 10 × rNTPs   5 μl
   - water   26.5 μl
   - DNA template   1 μl   (approx. 0.1–1.0 μg)
   - [^{32}P]UTP   5 μl   (equivalent to 12.5 μM)
   - RNA polymerase   1 μl   (10–20 units)
2. Incubate at 37°C for 1 h (bacteriophages T3 and T7)
             40°C for 1 h (bacteriophage SP6)
3. Optional: Remove template DNA by treatment with RNase-free DNase (1 unit) for 15 min at 37°C. Extract with an equal volume of phenol: chloroform (1:1 v/v) as described in *Protocol 10* and then remove unincorporated labelled nucleotides by chromatography on Sephadex (*Protocol 5*).

Protocol 11. *Continued*

4. Ethanol precipitate the RNA as described in *Protocol 5*. Redissolve the RNA in 100 µl DEPC-treated water.

B. Non-radioactive

Materials

As above, with the following changes:

- 10 × TB: 400 mM Tris–HCl, pH 7.5, 60 mM $MgCl_2$, 20 mM spermidine, 100 mM DTT, 100 mM NaCl
- 10 × dig–rNTP labelling mixture (10 × dig–rNTPs): 10 mM each rATP, rGTP, rCTP, 6.5 mM rUTP/3.5 mM dig–UTP, pH 7.5

or

- 10 × biotin rNTP labelling mixture (10 × bio–rNTPs): 10 mM each rATP, rGTP, rCTP, 6.5 mM rUTP/3.5 mM bio-UTP, pH 7.5.

Method

1. Add the following components to a microfuge tube on ice:

   - 10 × TB   2 µl
   - Water   12 µl
   - RNasin   1 µl
   - linearized template DNA   1 µl (1 µg)
   - 10 × dig–rNTPs or   2 µl
     10 × bio–rNTPs
   - RNA polymerase   2 µl (40 units)

2. Incubate at 37°C for 2 h (bacteriophages T3 and T7)
              40°C for 2 h (bacteriophage SP6)

3. Stop the reaction by addition of 0.2 M EDTA, pH 8.0 and ethanol precipitate the RNA as described in *Protocol 5*, except use lithium chloride instead of sodium acetate. Re-dissolve the RNA in 100 µl of DEPC-treated water. The efficiency of label incorporation is determined as described in Section 3.3.

The following points should be noted about the method in *Protocol 11*.

(a) The DNA template must be added after the transcription buffer has been diluted, as the 10 × transcription buffer contains spermidine at a concentration of 10 mM. Spermidine concentrations greater than 4 mM may cause the DNA to precipitate.

(b) The incorporation of bio–UTP into RNA by SP6 polymerase is very low, so T7 or T3 polymerases should be used.

(c) Phenol extraction of digoxigenin-labelled RNA following the transcription reaction gives poor yields, and should be omitted. It is attributed to partitioning of the labelled RNA into the organic phase (18).

If very high specific activity probes are required, for techniques such as *in situ* or filter hybridization and full-length trancription is not vital, all the 'cold' rUTP can be omitted and the reaction volume reduced to as little as $10\,\mu l$. Maximal specific activity of the probe will be achieved in a $20\,\mu l$ labelling reaction using $100\,\mu Ci$ of $400\,Ci/mmol$ UTP. The final concentration of UTP is $12.5\,\mu M$ (27).

For *in situ* hybridization, where signal localization at the single-cell level is required, $[\alpha\text{-}^{35}S]UTP$ is probably the radiolabelled nucleotide of choice, since it gives better resolution than ^{32}P. A direct substitution for $[\alpha\text{-}^{35}S]UTP$ can be made in the labelling reaction.

Non-radioactively labelled RNA probes have been successfully used to detect mRNA in archival biopsies by *in situ* hybridization (J. C. Martinez-Montero [submitted], and see Volume I). These probes may be stored for more than 2 years at $-20\,°C$ with no degradation of the RNA, which is significantly longer than the half-life of ^{32}P- and ^{35}S-labelled probes.

5.5 Advantages and specialist uses of RNA probes

RNA probes have many advantages over double-stranded DNA probes labelled by nick-translation and random priming: lack of competition of probe/probe hybridization, high specific activity (1×10^9 d.p.m./μg), and readily defined length. They can substitute for DNA probes in all circumstances, and the higher thermal stability of RNA–RNA hybrids, over RNA–DNA hybrids, gives increased sensitivity. The ability to use RNase A to digest unhybridized (single-stranded) RNA probe results in extremely low background (29). Thus, excellent signal-to-noise ratios in both filter and *in situ* hybridization reactions can be obtained (ref. 30, and see Volume I).

More specialist procedures which have been developed for use with RNA probes include: RNase mapping, the positional mapping of termini of target molecules (31), *in vitro* translation (32), and as antisense RNA to specifically block gene expression (33).

6. Labelling of oligonucleotides for use as hybridization probes

6.1 Types and uses of oligonucleotide probes

Short tracts of single-stranded DNA, termed oligonucleotides, are now routinely synthesized chemically by highly automated machines (34). The ready availability of oligonucleotides of defined sequence has prompted their use in a number of different applications. Synthetic oligonucleotides have been used as DNA primers in DNA sequencing (35), as hybridization probes in Southern analysis to detect specific groups of genes, to isolate genes and even to detect point mutations in specific alleles (36–37) (see Chapter 10).

There are three types of oligonucleotide probe in common use: single oligonucleotides of defined sequence; pools of short oligonucleotides whose sequences are highly degenerate; and pools of longer oligonucleotides of lesser degeneracy.

Generally, single oligonucleotide probes of defined sequence are only synthesized to part of a sequence of a previously cloned segment of DNA, and they match their target sequence perfectly. They are generally 19–40 nucleotides long, and when hybridized under the correct stringency conditions can discriminate between their target sequence and other closely related sequences. Hence, amongst other applications, they have been used to screen cDNA and genomic DNA libraries for overlapping clones containing DNA which has previously been isolated and sequenced.

The sequences of degenerate pools of oligonucleotides are normally deduced, using the genetic code, from short stretches of amino-acid sequences determined by the sequencing of small quantities (picomolar) of highly purified proteins (38). The degeneracy of the genetic code usually requires the synthesis of several oligonucleotides to represent all possible ways to code for a given short sequence of amino acids. The pools of degenerate nucleotides may then be used as hybridization probes to select a series of candidate clones en route to the isolation of a specific target gene. This approach has been applied successfully to the isolation of both cDNA (39) and genomic DNA clones (40). Oligonucleotides in degenerate pools are normally 17–20 nucleotides in length, long enough to hybridize specifically to the gene of interest, but short enough not to form internally mismatched hybrids to non-homologous sequences that can be detected by hybridization.

Pools of longer oligonucleotides (30–70 bp) of lesser degeneracy (that is, that contain only a subset of the possible codons at each position ['guessmers']) are successful because the detrimental effects of mismatches are outweighed by the increased stability of hybrids formed by longer oligonucleotides.

6.2 5′-end-labelling

Oligonucleotides are synthesized without a phosphate group at their 5′ termini and are therefore usually labelled by phosphorylation of 5′ termini with [γ-^{32}P]ATP, a reaction catalysed by bacteriophage T4 polynucleotide kinase (see *Protocol 12* and *Figure 5*). This results in a single labelled nucleotide per molecule (41). The specific activities achievable by such an end-labelling technique are significantly lower than those obtained by the uniform labelling techniques of nick-translation, random primer labelling, PCR and *in vitro* transcription. Indeed, the specific activity of the oligoprobe is only equal to that of the labelled nucleotide. For this reason ^{32}P-labelled nucleotides are most frequently used for end-labelling.

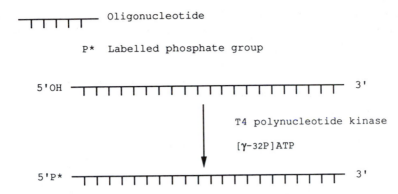

Figure 5. 5'-end-labelling of oligonucleotide using T4 polynucleotide kinase.

It is very important to separate the labelled oligonucleotides from the unincorporated [γ-^{32}P]ATP and to ensure that probe hybridization to the target sequence is not reduced by any contaminating unlabelled oligonucleotide. Both of these requirements can be achieved by (a) thin-layer chromatography (TLC), or (b) polyacrylamide gel electrophoresis in the presence of 7 M urea. The latter method is described in detail in another volume in this series (42).

To 5'-end-label oligonucleotides non-radioactively, they must be specifically synthesized with a 5' terminal amino group to which biotin can be readily attached using a chemical rather than an enzymatic reaction. This procedure is described by Chu and Orgel (43).

Protocol 12. 5'-end-labelling using T4 polynucleotide kinase

Materials

- Oligonucleotide (approx. 0.1 μg for a 19-mer)
- 10 × kinase buffer (10 × KB): 500 mM Tris–HCl, pH 7.6, 100 mM MgCl$_2$, 50 mM DTT, 1 mM spermidine
- T4 polynucleotide kinase: 4 units/μl
- [γ-^{32}P]ATP (specific activity 3000 Ci/mmol; 110 TBq/mmol, 10 μCi/μl)

Method

1. Add the following on ice to a 0.5-ml microcentrifuge tube in the order given.

   - Oligonucleotide   1 μl   (15 pmol ends)
   - Water to a **final** volume of   10 μl

Protocol 12. *Continued*

- 10 × KB   1 μl
- [γ-³²P]ATP*ᵃ*   1 μl   (approx. 150 μCi)
- T4 polynucleotide kinase   1 μl   (4 units)

Mix gently by slowly pipetting up and down. Centrifuge briefly at 12 000 g in a microfuge (2 sec) to collect contents at bottom of tube.

2. Incubate at 37°C for 45 min.

3. Remove unincorporated nucleotides by Sephadex G50 filtration, gel electrophoresis, HPLC, or TLC (42).

ᵃ It is necessary to dry down the label and re-dissolve it in 1 or 2 μl of water, so that the reaction volume can be kept to a minimum for subsequent purification of the oligoprobe by gel electrophoresis (42).

[γ-³⁵S]ATP can also be used as a donor in 5′-end-labelling, but higher enzyme concentrations are required for efficient reaction rates. Oligoprobes labelled with ³⁵S may be used for *in situ* hybridization.

6.3 3′-end-labelling

The enzyme terminal deoxynucleotidyl transferase (TdT) can be used in conjunction with ³²P, ³⁵S, or ³H radiolabelled nucleotides as well as with the non-radioactive labels, biotin and digoxigenin, to 3′-end-label oligonucleotides for use as hybridization probes. TdT catalyses the addition of one or more nucleotides on to the 3′-end of the oligonucleotide (44). The reaction requires the presence of a divalent cation and is therefore inhibited by chelating agents such as EDTA. In the presence of Mg^{2+} the enzyme catalyses the addition of only one or two nucleotide residues. However, replacement of Mg^{2+} with Co^{2+}, results in a more efficient reaction and the addition of more than two residues.

In order to incorporate a single radioactive end-label, a dideoxynucleoside 5′-triphosphate (usually [α-³²P]ddATP) is used (see *Figure 6*). This lacks a 3′-hydroxyl group resulting in termination of polymerization when just one residue has been added (45). A dig-11-ddUTP molecule will soon be available from Boehringer Mannheim allowing the addition of one defined non-radioactively labelled molecule per nucleotide.

For the addition of many labelled residues, the α-position of the phosphate or an internal residue of the base must be labelled. Probes should not be tailed with dTTP since the tail can hybridize to poly(A)⁺ sequences in mRNA giving rise to high background signals in *in situ* hybridization. Similarly, probes tailed with [α-³²P]dATP may hybridize to poly(T) regions in genomic DNA. The length of the tail (and the specific activity of the resultant probe) can be influenced by the relative molar concentrations of 3′ oligonucleotide

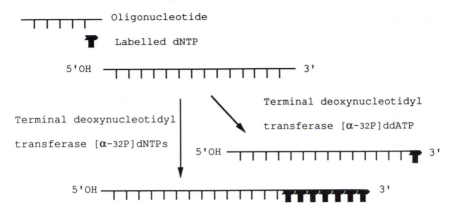

Figure 6. 3'-end-labelling of oligonucleotide using TdT.

ends and dNTPs. Using the conditions given in *Protocol 13*, about 60% of the [α-^{32}P]dCTP precursor is incorporated into the tail (approx. 100 residues in length). The incorporation of [α-^{35}S]dCTP is slightly lower so more enzyme and longer incubation times should be used.

Protocol 13. 3'-end-labelling using terminal deoxynucleotide transferase

A. Radioactive

Materials

- 5 × TdT tailing buffer: 5 mM $CoCl_2$, 500 mM sodium cacodylate, pH 7.0, 1 mM DTT
- [α-^{32}P]dCTP (specific activity 5000 Ci/mmol; 185 TBq/mmol)
- Oligonucleotide 20 ng/μl (for a 20-mer, this is equivalent to 3 pmol of ends)
- Terminal deoxynucleotidyl transferase (TdT): 5 units/μl

Caution: Cacodylate is a poisonous arsenic compound and should be handled with care. Contact with skin should be avoided.

Method

1. Add the following to a microfuge tube on ice:
   - 5 × TdT buffer   10 μl
   - Oligonucleotide   1 μl   (3 pmol ends)
   - Water   18 μl
   - [α^{32}-P]dCTP   20 μl (200 μCi)
   - TdT   1 μl   (5 units)

59

Protocol 13. *Continued*

2. Mix, and centrifuge briefly at 12 000 *g* in a microfuge to bring the contents to the bottom of the tube. Incubate the tube at 37°C for 45–60 min.

3. Stop the reaction by heating to 65°C for 5 min. Cool on ice.

4. Remove unincorporated nucleotides by gel electrophoresis, HPLC, or TLC (42).

B. Non-radioactive

Materials

As above, with the following changes:

For biotin:

- 5 × TdT tailing buffer: 5 mM $CoCl_2$, 700 mM sodium cacodylate, 0.5 mM DTT, 150 mM Tris–HCl, pH 6.8
- Bio-11-dUTP, 0.4 mmol
- Oligonucleotide (up to 1 μg: approx. 150 pmol ends of a 20-mer)
- TdT: 50 units/μl

Method

1. Add the following to a microfuge tube on ice:
  - 5 × TdT buffer   10 μl
  - Oligonucleotide   10 μl (150 pmol)
  - Bio-11-dUTP   1.75 μl
  - TdT   1 μl (50 units)
  - Water   27.25 μl

2. Mix, and centrifuge briefly at 12 000 *g* to bring the contents to the bottom of the tube. Incubate the tube at 37°C for 2 h.

3. Make up the volume to 100 μl and purify through a spun column, as described in *Protocol 5*. The labelled probe has up to three biotin molecules at the 3′-end. Store at −20°C.

For digoxigenin:

Materials

As above, with the following changes:

- 5 × TdT buffer: 1 M potassium cacodylate, 125 mM Tris–HCl, pH 6.6, 1.25 mg/ml bovine serum albumin)
- $CoCl_2$, 5 mM
- dATP, 2.5 mM

- dig-11-dUTP, 1 mM
- TdT: 55 units/μl
- Oligonucleotide (up to 1 μg: approx. 150 pmol ends of a 20-mer)

Method

1. Add the following to a microfuge tube on ice:
   - 5 × TdT buffer   10 μl
   - CoCl$_2$ 20 μl
   - dATP   3.5 μl
   - Oligonucleotide   10 μl (100 pmol)
   - dig-11-dUTP   1 μl
   - TdT   1 μl   (50 units)
   - Water   4.5 μl
2. Mix, and centrifuge briefly at 12 000 g to bring the contents to the bottom of the tube. Incubate the tube at 37°C for 15 min.
3. Make up the volume to 100 μl and purify through a spun column as described in *Protocol 5*. The oligoprobe has a mixed tail of labelled and unlabelled nucleotides. Store at −20°C.

The labelling methods given in *Protocol 13* are usually employed for short oligonucleotides (20–30 nucleotides in length). Longer oligonucleotides cloned into bacteriophage M13 vectors (approx. 70 nucleotides long or more), however, are almost always labelled in a primer–extension reaction using 'unversal' primer (46). Alternatively, longer probes can be prepared by annealing two separate oligonucleotides that contain complementary sequences at their 3′ termini. The protruding single-stranded regions can then be 'filled-in' using Klenow polymerase in the presence of labelled nucleotides. This primer extension procedure has been described in a previous volume of this series (42).

The specific activities of oligoprobes are much lower than those of uniformly labelled probes (approx. 5×10^6 d.p.m./μg) because usually only one labelled nucleotide is incorporated. The greatest sensitivity of hybridization will, therefore, be achieved using a cocktail of several oligonucleotides complementary to different regions of the target nucleic acid (47).

7. Conclusion

Plasmids have become the vectors of choice for molecular cloning because of their reliability and ease of handling. DNA fragments of potential interest, which are subcloned into plasmids, require transformation into a bacterial host for their propagation. We have described a simple, but efficient method for the introduction of supercoiled plasmid DNA into bacteria.

Of the various methods which have been developed for purifying recombinant bacterial plasmids, the procedure devised by Birnboim and Doly (10) is the most generally useful for both small- and large-scale work, and is described in Section 2.

We have summarized the variety of radioactive and non-radioactive labels available for use with nucleic acids, and described a number of enzymatic techniques which can be used to introduce them into nucleic-acid molecules.

Generally, non-radioactive probes are approximately as sensitive as ^{32}P-labelled probes, when the specific activity of the radioactive probe is 1–5 × 10^8 c.p.m./μg, but are usually less sensitive when compared with probes of 10^9 c.p.m./μg or greater. For most hybridization applications, any of the uniform labelling procedures (nick-translation, random priming, PCR, and 'phage polymerases) can be used successfully. For specialist techniques such as *in situ* hybridization of mRNA, however, RNA probes or labelled oligonucleotide 'cocktails' are often the probes of choice (see Volume I).

Acknowledgements

I would like to thank Lisa Ramshaw, John Loughlin, and Mark Evans for their critical reading of this manuscript and Sean McCarthy for preparing the figures. The original work here was supported by the Cancer Research Campaign, UK.

References

1. Cohen, S. N., Chang, A . C. Y., and Hsu, L. (1972). *Proc. Natl. Acad. Sci. USA,* **69,** 2110.
2. Yanisch-Perron, C., Vieira, J., and Messing, J. (1985). *Gene, 33,* 103.
3. Ullman, A., Jacob, F., and Monod, J. (1967). *J. Mol. Biol., 24,* 339.
4. Krieg, P. A. and Melton, D. A. (1987). *Methods in enzymology,* Vol. 155 (ed. R. Wu), p. 397. Academic Press, Orlando, Florida.
5. Weinstock, G. M., Rhys, C., Berman, M. L., Hampar, B., Jackson, D., Silhavy, *et al.* (1983). *Proc. Natl. Acad. Sci. USA,* **80,** 4432.
6. Mandel, M. and Higa, A. (1970). *J. Mol. Biol., 53,* 159.
7. Hanahan, D. (1983). *J. Mol. Biol., 166,* 557.
8. Clewell, D. B. (1972). *J. Bacteriol., 11,* 667.
9. Hardy, K. G. (ed.) (1987). *Plasmids: A practical approach.* IRL Press, Oxford.
10. Birnboim, H. C. and Doly, J. (1979). *Nucleic Acids Res., 7,* 1513.
11. Canter, C. R. and Schimmel, P. R. (1980). In *Biophysical Chemistry,* Part III, p. 1251. W. H. Freeman, San Francisco.
12. Maniatis, T., Fritsch, E. F., and Sambrook, J. (ed.) (1989). *Molecular cloning, a laboratory manual* (2nd edn). Cold Spring Harbor Press, Cold Spring Harbor, NY.

13. Polak, J. M. and McGee, J. O'D. (ed.) (1990). In situ *hybridisation, principles and practice: A practical approach*. IRL Press, Oxford.
14. Chan, V. T. W., Fleming, K. A., and McGee, J. O'D. (1985). *Nucleic Acids Res.*, **13**, 8083.
15. Forster, A. C., McInnes, J. L., Skingle, D. C., and Symons, R. H. (1985). *Nucleic Acids Res.*, **13**, 745.
16. Reisfed, A., Rothenberg, J. M., Bayer, E. A., and Wilcheck, M. (1987). *Biochem. Biophys. Res. Commun.*, **142**, 519.
17. Viscidi, R. P., Connelly, C. J., and Yolken, R. H. (1986). *J. Clin. Microbiol.*, **23**, 311.
18. Holtk, H. J. and Kessler, C. (1990). *Nucl. Acids Res.*, **18**, 5843.
19. Rigby, P. W. J., Dieckmann, M., Rhodes, C., and Berg, P. (1977). *J. Mol. Biol.*, **113**, 237.
20. Feinberg, A. P. and Vogelstein. B. (1983). *Analyt. Biochem.*, **132**, 6.
21. Feinberg, A. P. and Vogelstein, B. (1984). *Analyt. Biochem.*, **137**, 266.
22. Mullis, K., Faloona, F., Scharf, S., Saiki, R., Horn, G., and Erlich, H. (1986). *Cold Spring Harbor Symp. Quant. Biol.*, **51**, 263.
23. Saiki, R. K., Gelfland, D. H., Stoffel, S., Scharf, S. J., Higuchi, R., Horn, G. T., *et al.* (1988). *Science*, **239**, 487.
24. Schowalter, D. B. and Sommer, S. S. (1989). *Analyt. Biochem.*, **177**, 90.
25. Lion, T. and Haus, O. A. (1990). *Analyt. Biochem.*, **188**, 335.
26. Chamberlain, M. and Ryan, T. (1982). In *The enzymes* (3rd edn), Vol. 1 (ed. P. D. Boyer), p. 87. Academic Press, New York.
27. Little, P. F. R. and Jackson, I. J. (1987). In *DNA cloning*, Vol. III: *A practical approach* (ed. D. M. Glover), p. 1. IRL Press, Oxford.
28. Schenborn, E. T. and Mierendorf, R. C. (1985). *Nucleic Acids Res.*, **13**, 6223.
29. Melton, D., Krieg, P. A., Rebagliati, M. R., Maniatis, T., Zinn, D., and Green, M. R. (1984). *Nucleic Acids Res.*, **12**, 7035.
30. Cox, K. H., De Leon, D. V., Angerer, L. M., and Angerer, R. C. (1984). *Dev. Biol.*, **101**, 458.
31. Zinn, K., diMaio, D., and Maniatis, T. (1984). *Cell*, **34**, 865.
32. Krieg, P. A. and Melton, D. A. (1984). *Nucleic Acids Res.*, **12**, 7057.
33. Melton, D. A. (1985). *Proc. Natl. Acad. Sci. USA*, **82**, 144.
34. Itakura, K., Rossi, J. J., and Wallace, R. B. (1984). *Ann. Rev. Biochem.*, **53**, 323.
35. Smith, M. (1983). In *Methods of DNA and RNA sequencing* (ed. S. M. Weissman), p. 23. Praeger, New York.
36. Wallace, R. B., Schold, M., Johnson, M. J., Dembek, P., and Itakura, K. (1981). *Nucleic Acids Res.*, **9**, 3647.
37. Caruthers, M. H. (1985). *Science*, **230**, 281.
38. Hunkapiller, M. W., Lujan, E., Ostrander, F., and Hood, L. E. (1983). *Methods in enzymology*, Vol. 91 (ed. C. H. W. Hirs and S. W. Timashoff), p. 227. Academic Press, New York.
39. Sood, A. K., Pereira, D., and Weissman, S. M. (1981). *Proc. Natl. Acad. Sci. USA*, **78**, 616.
40. Jacobs, K., Schomaker, C., Rudersdorf, R., Neill, S. D., Kaufman, R. J., Mufson, A., *et al.* (1985). *Nature*, **313**, 806.
41. Connor, B. J., Reyes, A. A., Morin, C., Itakura, K., Teplitz., R. L., and Wallace R. B. (1983). *Proc. Natl. Acad. Sci. USA*, **80**, 278.

42. Thein, S. L. and Wallace, R. B. (1986). In *Human genetic diseases: A practical approach* (ed. K. E. Davies), p. 33. IRL Press, Oxford.
43. Chu, B. C. F. and Orgel, L. E. (1985). *DNA,* **4,** 327.
44. Bollum, F. J. (1974). In *The enzymes* (ed. P. D. Boyer), Vol. X, p. 145. Academic Press, London.
45. Yousaf, S. I., Carrol, A. R., and Clarke, B. E. (1984). *Gene,* **27,** 309.
46. Anderson, S. and Kingston, I. B. (1983). *Proc. Natl. Acad. Sci. USA,* **80,** 6838.
47. Pringle, J. H., Ruprai, A. K., Primrose, L., Keyte, J., Potter, L., Close, P., and Lauder, I. (1990). *J. Pathol.,* **162,** 197.

3

Molecular hybridization of nucleic acids

VICTOR T-W. CHAN

1. Introduction

Molecular hybridization is the reaction in which single-stranded target nucleic acid sequences re-anneal to complementary molecular probes to form double-stranded hybrid molecules. In general, the target sequences are part of a complex population of heterogeneous species of nucleic acids, and labelled probes must therefore search for and anneal to their complementary targets amongst this mixture. Molecular probes can be labelled either with radio-isotopes or with non-radioactive reporter molecules (see Chapter 2). Both DNA and RNA molecules can be radioactively and non-radioactively labelled, either enzymatically or chemically, and used as hybridization probes. DNA probes are usually double-stranded, so they require denaturation to single-stranded molecules before hybridization. This is most simply achieved by boiling. Conversely, RNA probes are usually single-stranded, and denaturation is not theoretically necessary. However, it is advantageous because RNA molecules assume secondary structures; denaturation therefore increases the proportion of fragments accessible for hybridization. In single-phase (that is, solution) hybridization, the optimal temperature for renaturation of nucleic acids is approximately 25 °C below the melting temperature (T_m) of the hybrid molecules. The T_m of a duplex nucleic acid molecule, which is defined as the temperature at which 50% of the duplexes are dissociated to single-stranded molecules, depends on guanine–cytosine (G–C) content (expressed as a percentage), the length of the duplex [in base pairs (bp)], the concentration of monovalent (for example, Na^+) cation (M), and the concentration of forma-mide (%) in the reaction mix. By using the following formula, the T_m of the hybrid molecules with known G–C content and size in a particular solution can be estimated.

$$T_m = 81.5 + 16.6 \log M + 0.41 \text{ (% G–C)} - 0.72 \text{ (% formamide)} - 500/\text{size in bp (1)}.$$

It should be emphasized that this is only an estimation. The exact figure must be determined experimentally because it not only depends on the G–C proportion but also on its distribution. An additional correction is applicable to mismatched hybrids. In general, 1% mismatch reduces the T_m by 0.5–1.5 °C. The stringency of hybridization and post-hybridization washes determines the degree to which mismatched hybrids are permitted to form and remain respectively. Hybridization is usually performed at relatively low stringency (such as 25 °C below the T_m of perfectly matched hybrids), allowing rapid formation of hybrid molecules with only partial sequence homology. Higher stringency is then achieved by post-hybridization washing to the required level of stringency.

In practice, hybridization reactions are often carried out in two-phase systems such as on filters or *in situ*. In two-phase hybridization reactions, the kinetics of hybrid formation as well as the T_m of the hybrid molecules are different from that in single-phase hybridization reactions. Therefore, the hybridization time and stringency for hybridization and post-hybridization washes must be determined empirically (see Volume I). Further information regarding the principles of molecular hybridization are given in another volume in this series (1).

1.1 The application of blot hybridization to the study of human disease

The presence of abnormal (altered or exogeneous) nucleic acid sequences can be determined in this way in any clinical sample from which nucleic acid can be extracted (see Chapter 1). Blots prepared from samples fractionated by gel electrophoresis not only demonstrate the presence of the target sequences but also changes in molecular size, indicating structural alterations. Direct application of nucleic acids on to membrane filters (dot blots) can be used to indicate the presence or absence of target sequences but gives no structural information. Northern (RNA) blots are usually used to characterize changes in levels of gene expression. Furthermore, this technique can be used to demonstrate alterations in size (or splicing) of mRNA transcripts. Changes either in size or level of expression of mRNA transcripts may significantly alter either the amount of protein product, or its biological activity. One of the best examples of this is retinoblastoma (Rb) gene mutation which either produces abnormal mRNA transcripts or abolishes normal expression of the gene. Inactivation of this gene is implicated in the aetiology of many human cancers. Similarly, enhanced expression of oncogenes or expression of altered oncogene products are also implicated in many human cancers and may be studied in a similar way. Southern (DNA) blots are usually used to study the structure of the gene locus of interest. By studying restriction fragment length polymorphisms (RFLPs), changes in gene loci can be identified provided that suitable restriction enzymes are available (see Chapter 10). Furthermore, by using a pair of isoschizomers which recognize methylated and non-methylated nucleotides (for example, 5-methylcytosine residues), the level of methylation or changes in the methylation pattern of a gene can be determined. In many genetic diseases (such as sickle-cell anaemia, thalassaemias, cystic fibrosis, Duchenne muscular destrophy, etc.), structural changes in the appropriate gene loci can be identified on Southern blots. Furthermore, activation of oncogenes by gene rearrangement, point mutation, or gene amplification, and inactivation of tumour suppression genes by DNA rearrangements, point mutations or allelic deletions can also be demonstrated in this way (see Chapter 10).

The presence of infectious (microbial or viral) agents can be demonstrated by the presence of their genomes on DNA blots, or by the presence of RNA transcripts encoded by their genomes on RNA blots. Furthermore, the specific strain of these infectious agents can be identified by using segments of its genome, which are highly specific for that particular strain, as molecular probes in the hybridization reactions (see Chapter 8).

2. Blot hybridization

There are two approaches to nucleic acid blotting:

(a) direct spotting of nucleic acid solutions on to a filter matrix (dot or slot blots);

(b) transfer of nucleic acids after fractionation from an agarose gel to a membrane filter (Southern or Northern blots).

For DNA, the latter approach involves digestion with restriction enzymes, fractionation by gel electrophoresis, and transfer from the agarose gel to a sheet of membrane filter by capillary flow of a buffer with high ionic strength (for example, $20 \times$ SSC). Since this technique was developed by E. M. Southern, the blots prepared in this way are called Southern blots (2).

Native RNAs, with a significant degree of secondary structure, do not bind efficiently to membrane filters. However, denatured (completely single-stranded) RNA can be efficiently immobilized. RNA samples are therefore first denatured, fractionated by gel electrophoresis, and transferred from the agarose gel to a sheet of membrane filter similar to that used for preparing Southern blots. This procedure is designated Northern blotting or Northern transfer (3, 4) by comparison with Southern transfer for DNA samples.

After DNA or RNA samples have been transferred to (or applied on to) filters, they must be permanently fixed so that they do not dissociate from the filter during hybridization and post-hybridization washes. This is achieved by either baking nitrocellulose at 80°C under vacuum, or by cross-linking to nylon filters by UV irradiation.

In general, blot hybridization reactions are driven by the nucleic acids immobilized on the filters. However, they can also be driven by high probe concentration in the hybridization buffer, or by the addition of high-molecular-weight anionic polymer (such as dextran sulfate) to the hybridization mix (5). However, the kinetics of hybridization are difficult to predict from theoretical considerations, partly because the exact concentration of the immobilized nucleic acid and its accessibility for hybridization are unknown. In practice, overnight (16 h) hybridization in the presence of 10% dextran sulfate is sufficient for the detection of single-copy genes, as well as low abundance mRNA.

2.1 DNA blot hybridization

DNA blot hybridization has two components: preparation of DNA blots followed by hybridization. These are discussed in turn below.

2.1.1 Preparation of Southern and DNA dot blots

Many protocols, as well as many modifications of these protocols, have been described for the preparation of DNA blots. However, some of them require special apparatus or membrane filters. The procedures described in *Protocols 1* and *2* combine simplicity, generality, and efficiency. More importantly, they are highly reproducible and sensitive (6).

Protocol 1. Preparation of Southern blots

Materials

- 10 × TBE buffer: 0.9 M Tris base, 0.9 M boric acid, 20 mM EDTA. Adjust the pH to 8.1–8.2 by the addition of solid boric acid.
- Ethidium bromide, 10 mg/ml
- Loading buffer: 0.25% bromophenol blue, 40% sucrose in 1 × TBE
- 0.25 M HCl
- Denaturing buffer: 1.5 M NaCl, 0.5 M NaOH
- Neutralizing buffer: 0.5 M Tris–HCl, pH 7.0, 3 M NaCl
- 20 × standard saline citrate (SSC): 3 M NaCl, 0.3 M sodium citrate, pH 7.0
- Membrane filters (nitrocellulose or nylon)
- Agarose

Method

1. Set up restriction digestion of DNA samples following the instruction of the suppliers of the restriction enzymes. In general, overnight digestion with five units of enzyme per microgram of DNA is sufficient

to give complete digestion. However, it is important to use the correct buffer to avoid non-specific star activity (see Chapter 10 for details).

2. Fractionate the DNA digests by agarose gel electrophoresis (with molecular weight markers) at 1–2 V/cm. The length of the gel depends on the range of fragment size. If the difference is large, a long gel (20 cm) should be used. Conversely, a single fragment, or fragments of similar size, can be analysed on a relatively short gel (10 cm). In most experiments, 0.8–1.0% agarose can be used to separate DNA fragments of different size (1–15 kb). If the fragments are small (400–800 bp), a higher concentration of agarose (1.2–1.4%) should be used. By contrast, lower concentrations of agarose (0.5–0.7%) should be used for larger fragments (15–30 kb). For details of gel electrophoresis of DNA, see Chapter 1.

3. After electrophoresis, photograph the gel with a Polaroid camera. It is a good idea to photograph a ruler alongside the gel so that the mobility of the marker fragments can be easily determined.

4. Transfer the gel to a plastic box, upside down and rinse it in water several times. Take care not to drop or break the gel.

5. Treat the gel with 0.25 M HCl for 20–35 min and rinse it several times in water.[a]

6. Denature the DNA fragments by incubating the gel in denaturing buffer for 35 min with gentle shaking or rocking.

7. Repeat the denaturation with fresh buffer for 20 min and rinse the gel in water several times.

8. Neutralize the gel in two changes of neutralizing buffer for 15 min each. The buffer should be changed after the first wash.

9. During the treatment of the gel, cut a piece of filter (nitrocellulose or nylon) about the same size as the gel.

10. Wet the filter in water, then saturate it with 20 × SSC. Set up the transfer apparatus as shown in *Figure 1*. Cover the transfer apparatus with a piece of plastic wrap.

11. Cut an opening in the plastic wrap which is *exactly the same size* as the gel and cover the exposed area with a small volume of 20 × SSC. This will reduce the possibility of trapping any air bubbles between the gel and the wicks.

12. Layer the gel on to the opening and cover it with a small volume of 20 × SSC.

13. Layer the membrane filter on to the gel, making sure that no air-bubbles are trapped between the gel and the filter.

Protocol 1. *Continued*

14. Layer two sheets of Whatman 3 MM filter paper on to the membrane filter, followed by a stack of absorbent paper (or blotting paper).

15. Put a piece of glass and a weight of approx. 500 g on top of the absorbing paper.

16. Blot the gel for 12–16 h.

17. After blotting, transfer the membrane filter together with gel on to a piece of dry filter paper with the gel side up. Use a soft pencil to mark the position of the gel and the wells on the filter. Peel off the gel, and stain it in ethidium bromide (0.5 μg/ml in water) to check the efficiency of transfer.

18. Rinse the filter in 5 × SSC several times,[b] transfer it on to a piece of dry filter paper, and trim off the edges.

19. Bake nitrocellulose filters at 80°C under vacuum for 2 h or irradiate nylon filters with UV light. DNA fragments can be crosslinked to nylon filters using either a commercial UV crosslinking machine or by irradiating the filter (with DNA side down) on a UV transilluminator for 3–5 min.

20. The filter is ready for hybridization or can be stored at 4°C. Nitrocellulose filters should be stored under vacuum.

---

[a] The time required for acid depurination depends on the concentration of agarose, the thickness of the gel, and the size of the fragments of interest. However, it is important to ensure that DNA fragments are not over-depurinated.
[b] It is important to remove residual agarose from the filters after transfer.

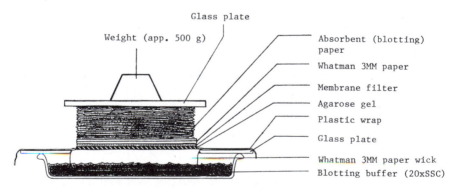

Figure 1. Southern transfer of DNA from agarose gel to membrane filter.

Protocol 2. Preparation of DNA dot blots

Materials
(see also *Protocol 1*)

- TE buffer: 10 mM Tris–HCl, 1 mM EDTA, pH 8.0
- 20 × SSC
- Membrane filters

Method

1. Sonicate genomic DNA at approx. 70 W for 1–2 min. The size of the DNA fragments should be 5–10 kb. (Check by agarose gel electrophoresis; see Chapter 1.)

2. Dilute the DNA to the desired concentration in TE buffer (see Chapter 1). In general, 2.5 μg of total genomic DNA in 5 μl should be used.

3. Denature the DNA samples by boiling in a water bath for 10 min, then chill on ice for 5 min and spin briefly at 12 000 g to collect the contents at the bottom of the tube.

4. During denaturation, cut an appropriately sized piece of membrane filter, wet it in distilled water, then saturate it in 20 × SSC.

5. Layer the membrane filter on a sheet of filter paper pre-saturated with 20 × SSC.

6. Spot the samples on to the filter directly.[a]

7. Transfer the filter to a sheet of dry filter paper and rinse in 5 × SSC for 5 min.

8. Fix the DNA to the membrane filter by baking at 80°C under vacuum (nitrocellulose), or by UV crosslinking for 3–5 min (nylon). The filter is now ready for hybridization or can be stored at 4°C. Nitrocellulose filters should be stored under vacuum.

[a] The filters should remain wet (but not too wet) throughout the spotting of DNA samples. DNA should be diluted in a buffer of low ionic strength (such as TE) and free of acetate ions. Alternatively, commercial filtration manifolds can be used to prepare DNA dot (slot) blots. The supplier's instruction for the use of the apparatus should be followed.

2.1.2 Hybridization of DNA blots

DNA blot hybridization can be performed under many different conditions, and the conditions used depend mainly on personal preference. In general, the rate of nucleic acid re-association at high temperature (65°C) is higher than that at low temperature (42°C) in the presence of formamide provided that the overall stringency is the same. For this reason, the procedure described in *Protocol 3* uses high temperature hybridization.

Protocol 3. Hybridization of DNA blots

Materials
(see also *Protocol 1*)

- Pre-hybridization buffer: 5 × SSPE, 5 × Denhardt's solution, 100 μg/ml denatured sonicated salmon sperm DNA, and 0.1% (v/v) SDS. The concentration of SDS should be increased to 1% (v/v) if nylon filters are used.
- Hybridization mixture: 5 × SSPE, 5 × Denhardt's solution, 100 μg/ml denatured sonicated salmon sperm DNA, 0.1% (v/v) SDS, 10% (w/v) dextran sulfate and denatured labelled probe.[a] The concentration of SDS should be increased to 1% (v/v) if nylon filters are used.
- Wash buffer 1: 2 × SSC, 0.1% SDS
- Wash buffer 2: 0.25 × SSC, 0.1% SDS
- 20 × SSC
- 20 × SSPE: 3 M NaCl, 0.3 M sodium phosphate, 40 mM EDTA, pH 7.2
- 100 × Denhardt's solution: 2% (w/v) BSA, 2% (w/v) Ficoll 400, 2% (w/v) polyvinylpyrrolidone (PVP) 360

Method

1. Wet the filter containing the immobilized DNA fragments (see *Protocols 1* and *2*) in water and place it in a hybridization bag.
2. Add pre-hybridization buffer to the bag (use 10 ml of buffer per 100 cm^2 of filter), squeeze all the air-bubbles out, and seal the bag with a heat sealer. Pre-hybridize the filter at 65°C for 2–4 h between two pieces of glass.
3. Cut off a corner of the bag and squeeze out the pre-hybridization buffer.
4. Add hybridization mixture containing denatured probe to the filter (use 4 ml of buffer per 100 cm^2 of filter). The probe can be denatured by incubating it in a boiling water bath for 10 min.
5. Squeeze all the air-bubbles out, re-seal the bag, and hybridize the filter at 65°C for 16 h between two pieces of glass.[b]
6. After hybridization, open the hybridization bag carefully without damaging the filter, and squeeze out the hybridization mixture. The mixture can be re-used for another hybridization reaction. If radioactive probes are used, great care must be taken to ensure that the radioactivity is well contained and properly disposed.
7. Rinse the filter several times in wash buffer 1 and wash in the same buffer for 3 × 5 min at room temperature.
8. Wash the filter in three changes of wash buffer 2 for 5 min each at room temperature.

9. Wash the filter in three changes of wash buffer 2 for 30 min each at 55–65°C.[c]

10. Briefly blot the filter dry and wrap it in plastic wrap.

11. Set up autoradiography with an intensifying screen in an autoradiographic cassette if radioactive probes are used. The intensifying screen should be placed on the bottom, the X-ray film in the middle and the blot on the top with the DNA side facing the film (see *Protocol 8*). If non-radioactive probes are used, the filters must be treated with blocking buffer to block background staining by the detection reagents (see *Protocol 9*). Alternatively, the filter can be stored damp at 4°C.

[a] If radioactive probes are used, they should be labelled to a specific activity of approx. 1×10^9 c.p.m./μg (see Chapter 2). In general, 10^6–10^7 c.p.m. per millilitre of hybridization buffer is sufficient. The kinetics of re-association of probes labelled with non-radioactive reporter molecules is slightly different from ^{32}P-labelled probes: a higher probe concentration is required for optimal sensitivity. The concentration of probe used can be calculated from the following formula:

$$A = B/50 \times 1/C \times 16/D$$

where A = concentration of probe required (μg/ml)
    B = the complexity of the probe (see ref. 1)
    C = concentration of dextran sulfate (%)
    D = hybridization time (in hours)
In general 50–100 ng/ml is sufficient for nick translated probes.

[b] If radiolabelled probe is used, the bag should be sealed in a second bag to prevent leakage.
[c] The stringency of the final washes can be increased by decreasing the concentration of SSC to $0.1 \times$ (instead of $0.25 \times$). This does not usually affect the sensitivity.

2.2 RNA blot hybridization

RNA blot hybridization involves hybridization of immobilized RNA (either total RNA or mRNA) with labelled DNA or RNA probes. The following two sections describe first the preparation of RNA blots followed by their hybridization to DNA probes.

2.2.1 Preparation of Northern and RNA dot blots

Native RNAs have significant secondary structure and therefore do not bind well to solid matrices (membrane filters). Efficient binding requires complete denaturation. Similarly, formation of secondary structure interferes with the mobility of RNA in gels and therefore their size cannot be accurately determined without denaturation. RNAs can be denatured either with glyoxal and dimethylsulfoxide (DMSO) (7) or formaldehyde and formamide (8). These two methods are equally efficient. However, the formaldehyde gel is relatively simpler and more reliable than the glyoxal method, and therefore it is more commonly used. The glyoxal/DMSO and formaldehyde/formamide methods are detailed in *Protocols 4* and *5* respectively.

Protocol 4. Preparation of Northern blots of RNA samples denatured with glyoxal and DMSO

Materials
(see also *Protocol 10*, Chapter 1)

All reagents should be RNase-free or DEPC-treated.

- 10 × Electrophoresis buffer: 100 mM sodium phosphate, pH 7.0
- Denaturing mixture: 3 ml 6 M glyoxal, 7.5 ml DMSO, 1.5 ml 10 × electrophoresis buffer. 12 μl of denaturing mix is added to 3 μl of RNA solution (containing up to 20 μg of RNA).
- Loading buffer: 50% of glycerol, 0.25% bromophenol blue in 1 × electrophoresis buffer
- Agarose

Method

1. Denature the RNA by incubation in denaturing mixture at 50°C for 1 h.
2. Fractionate the RNA in an agarose gel in 1 × electrophoresis buffer. In general, the length of gel required is 10–15 cm. 1.1% (w/v) agarose is generally adequate. If the RNA molecules are small (1 kb or smaller), 1.3–1.5% agarose should be used. For details of this gel system, see *Protocol 10*, Chapter 1.
3. After electrophoresis, cut the lanes containing the molecular weight markers from the gel and stain them with ethidium bromide (0.5 μg/ml in 25 mM Tris–HCl, pH 9.0) for 30–60 min.
4. Align a ruler with the gel on a UV transilluminator and photograph the gel and ruler with a Polaroid camera.
5. Gloxylated RNA can be transferred immediately after electrophoresis from an agarose gel to a membrane filter without any pre-treatment. In fact, any pre-treatment of the gel is detrimental. The set-up of blotting for RNA transfer is exactly the same as for DNA transfer (see *Protocol 1*). Before use, the membrane filter should be pre-saturated with 20 × SSC, which is also used as blotting buffer. Transfer is usually complete in 16 h.
6. After blotting, transfer the membrane filter with the gel to a piece of dry filter paper with the gel side up. Use a soft pencil to mark the positions of the gel and the wells on the filter.
7. Peel off the gel and stain it with ethidium bromide (0.5 μg/ml in 25 mM Tris–HCl, pH 9.0) to determine the efficiency of transfer.
8. Rinse the filter in 20 × SSC several times,[a] then trim off the edges.
9. Bake nitrocellulose filters at 80°C for 2 h under vacuum, or irradiate nylon

filters with UV light for 3–5 min on a UV transilluminator with the RNA side down. The filter is now ready for hybridization. Alternatively, it can be stored at 4°C. Nitrocellulose should be stored under vacuum.

[a] It is important to wash the filter with 20 × SSC after transfer to remove residual agarose. If desired, RNA can be washed off the filter using buffer of lower ionic strength before they are permanently fixed.

Protocol 5. Preparation of Northern blots from formaldehyde gels

Materials
(see also *Protocol 11*, Chapter 1)

All reagents should be RNase-free or DEPC-treated.

- Formamide, stored at $-70°C$ in small aliquots (see Chapter 1).
- Formaldehyde, 37% solution in water (see Chapter 1)
- 10 × Electrophoresis buffer: 0.2 M 3-(N-morpholeno)propanesulphonic acid (MOPS), 75 mM sodium acetate, 20 mM EDTA, pH 7.0
- Neutralizing buffer: 0.25 M Tris–HCl, pH 7.5, 3 M NaCl
- Loading buffer: 50% glycerol, 0.25% bromophenol blue in 1 × electrophoresis buffer

Method

1. Dissolve the RNA (20 μg) in 3.3 μl of water and add 1.5 μl of 10 × electrophoresis buffer, 7.5 μl of formamide and 2.7 μl of formaldehyde.
2. Denature the RNA at 65°C for 15 min and fractionate on an agarose gel containing 2.2 M of formaldehyde. In general, the length of the gel should be 10–15 cm and 1.1% (w/v) agarose can be used to separate most RNA molecules. If the RNA molecules are small (1 kb or smaller), 1.3–1.5% (w/v) agarose should be used. For details of formaldehyde gel electrophoresis, see *Protocol 11*, Chapter 1.
3. After electrophoresis, cut the lanes containing the molecular weight markers from the gel and stain them with ethidium bromide (0.5 μg/ml in water) for 30–60 min.
4. Align a ruler with the gel on a UV transilluminator and photograph both gel and ruler with a Polaroid camera.
5. Rinse the gel in water several times to remove the formaldehyde. Alternatively, the gel can be soaked in 0.05 M NaOH for 10–20 min to partially hydrolyse the RNA and thus improve the efficiency of transfer.[a]
6. After NaOH treatment, rinse the gel in water several times, then wash in neutralizing buffer for 30–45 min.

Protocol 5. *Continued*

7. Set up the blotting of the gel as described for DNA transfer (see *Protocol 1*). Before use, the membrane filter should be pre-saturated with 20 × SSC, which is also used as blotting buffer. The transfer is usually complete in 16 h.

8. Continue from step 6 of *Protocol 4*.

ᵃ The time required for NaOH treatment depends on the concentration of agarose, thickness of the gel and the size of the RNA molecules of interest. However, it is important to ensure that the RNA molecules are not over-hydrolysed.

Protocol 6. Preparation of RNA dot blots

Materials
(see *Protocols 10* and *11*, Chapter 1 for details of preparation)
All reagents should be RNase-free or DEPC-treated.

- 6 M glyoxal, stored at −20°C in small aliquots
- Formaldehyde, 37% solution in water
- Formamide, stored at −70°C in small aliquots
- 20 × SSC (see *Protocol 1*)

Method

1. Cut a piece of membrane filter, wet it in water, and then saturate it with 20 × SSC.

2. Denature the RNA samples with either glyoxal or formaldehyde and formamide.

    (a) For RNA denaturation with glyoxal, add the following:
    - 3 μl RNA solution (up to 20 μg)
    - 2 μl 6 M glyoxal
    - 5 μl 20 × SSC

    Heat the samples at 50°C for 1 h.

    (b) For RNA denaturation with formaldehyde and formamide, add the following:
    - 2 μl RNA solution (up to 20 μg)
    - 5 μl formamide
    - 2 μl formaldehyde
    - 1 μl 20 × SSC

    Heat the samples at 65°C for 15 min.

3. Chill the denatured RNA samples on ice for 5 min then spin briefly at 12 000 *g* to collect the contents at the bottom of the tube.

4. Layer the membrane filter on a sheet of filter paper pre-saturated with 20 × SSC.

5. Spot the denatured RNA samples on to the filter directly in two equal portions.[a]

6. Transfer the filter to a sheet of dry filter paper, air-dry briefly and rinse it several times in 20 × SSC.

7. Fix the RNA to the filter by baking at 80°C for 2 h under vacuum (nitrocellulose), or by UV irradiation on a UV transilluminator for 3–5 min, with the RNA side down (nylon). The filter is ready for hybridization, or can be stored at 4°C. Nitrocellulose should be stored under vacuum.

[a] The filter should remain wet (but not too wet) throughout the spotting of the RNA samples. Alternatively, commercial filtration manifolds can be used to prepare dot or slot blots. The supplier's instructions for the use of these filtration manifolds should be followed.

2.2.2 Hybridization of RNA blots

RNA blots are usually hybridized at 42°C in the presence of 50% formamide in hybridization buffer. Modifications have been made to this procedure in which the hybridization temperature is increased and the concentration of formamide decreased to balance the stringency in the hybridization reaction (for example, 48°C with 40% formamide). This is because the rate of nucleic acid re-association is higher at higher temperatures (for the same stringency). However, RNA degrades at high temperature, and thus dissociates from the membrane filter. Therefore, the efficiency of hybridization must be balanced against the stability of the RNA. *Protocol 7* uses the standard hybridization temperature of 42°C with 50% formamide.

Protocol 7. Hybridization of RNA blots

Materials
(see also *Protocol 3*)

All reagents should be RNase-free or DEPC-treated.

• 25 mM Tris–HCl, pH 8.5

• Prehybridization buffer: 5 × SSPE, 5 × Denhardt's solution, 100 μg/ml denatured sonicated salmon sperm DNA, 0.1% SDS, and 50% formamide. The concentration of SDS should be increased to 1% (v/v) if nylon filters are used.

Protocol 7. *Continued*

- Hybridization mixture: 5 × SSPE, 5 × Denhardt's solution, 100 μg/ml denatured sonicated salmon sperm DNA, 0.1% SDS, 50% formamide, 10% dextran sulfate, and denatured labelled probe (see *Protocol 3*). The concentration of SDS should be increased to 1% (v/v) if nylon filters are used.
- Wash buffer 1:2 × SSC, 0.1% SDS
- Wash buffer 2:0.25 × SSC, 0.1% SDS
- 20 × SSC: (see *Protocol 1*)
- 20 × SSPE: (see *Protocol 3*)
- 100 × Denhardt's solution: (see *Protocol 3*)

Method

1. Wet the filter in sterile distilled DEPC-treated water. Filters containing glyoxalated RNA must be heat-treated to remove the glyoxal adducts from the RNA molecules. Those containing RNA denatured with formaldehyde and formamide can be hybridized directly. Glyoxal adducts can be removed by heating the filters to 80°C in 25 mM Tris–HCl, pH 8.5 for 10 min and then cooling slowly to room temperature.

2. Place the filter in a plastic bag, add pre-hybridization buffer and seal the bag. Use 10 ml of buffer per 100 cm² of filter.

3. Squeeze all the air-bubbles out, seal the bag and pre-hybridize the filter at 42°C between two pieces of glass for 2–4 h.

4. Cut a corner off the plastic bag, squeeze out the pre-hybridization buffer and replace it with hybridization mixture containing denatured probe. Use 4 ml of mix per 100 cm² of filter. The probe can be denatured simply by boiling it for 10 min prior to addition to the hybridization mixture.

5. Squeeze out all the air bubbles, re-seal the bag and hybridize the filter at 42°C for 16–40 h between two pieces of glass.[a]

6. After hybridization, open the bag carefully without damaging the filter and squeeze out the hybridization mixture. This can be re-used for another hybridization reaction. If radioactive probes are used, great care should be taken to make sure that the radioactivity is well contained and properly disposed.

7. Rinse the filter in wash buffer 1 several times then wash it in three further changes of wash buffer 1 for 5 min each at room temperature.

8. Wash the filter in three changes of wash buffer 2 for 5 min each at room temperature and then in a further three changes of wash buffer 2 for 30 min each at 45–55°C.[b]

9. Briefly blot the filter dry and wrap it in plastic wrap.

10. Detect the hybridized probe by either autoradiography (radiolabelled probes) or histochemical means (non-isotopic probes) (see *Protocols 8 and 9*).

a If radiolabelled probe is used, the bag should be sealed in a second bag to prevent leaks.

b The stringency of the final washes can be increased by decreasing the concentration of SSC to 0.05 × (instead of 0.1 ×). This usually does not affect the sensitivity because the T_m of DNA–RNA and RNA–RNA hybrids is higher than that of DNA–DNA hybrids under the same conditions.

2.3 Detection of hybridized probes

After hybridization, the hybridized probes require detection to indicate the location (or presence) of target sequences. Both radioactive and non-radioactive DNA or RNA probes can be used. The detection of radioactive probes essentially involves the detection of their isotopic labels by auto-radiography. This is generally simple although long exposure times may be required. On the other hand, the detection of non-radioactive probes is more complicated and labour-intensive. This is because the reporter molecules require immuno/affinity chemical techniques for their detection. They are, however, safer to use and detection can generally be performed more quickly.

2.3.1 Detection of radiolabelled probes

For blot hybridization, the most commonly used radioisotope is ^{32}P which is a hard β emitter. The β particles emitted by ^{32}P can be recorded directly on X-ray film. However, the sensitivity can be significantly increased by using an intensifying screen which converts the energy of the β particles into fluorescence.

Protocol 8. Autoradiographic probe detection

Materials
- Autoradiographic cassette
- X-ray film (for example, Kodak X-Omat AR or equivalent)
- Intensifying screens (for example, DuPont Cronex Lightning-Plus or equivalent)

Method
1. After hybridization, briefly blot the filter dry and wrap it in plastic wrap. It is important to keep the filter damp if it has to be reprobed.*a*
2. In a dark room, place an intensifying screen in an autoradiographic cassette. Layer the X-ray film on to the intensifying screen and place the hybridized blot on to the film with the nucleic acid side facing the film. If two X-ray

Protocol 8. *Continued*

films are used, the blot must be taped on to the first film and the second film placed on to the blot followed by another intensifying screen.[b]

3. Close the cassette, wrap it in aluminium foil and expose the film(s) at −70°C.

4. Develop the film in an automatic X-ray film processor. Overnight exposure is usually sufficient to detect single-copy genes on Southern blots and low abundance mRNA in a poly(A)$^+$ RNA preparation on Northern blots. If two films are used, develop the first film after overnight exposure. The time required for optimal signal-to-noise ratio can be estimated from the intensity of the signal and background of the autoradiogram.

---

[a] Since ^{32}P is a hard β particle emitter, it must be handled carefully and individual exposure monitored. Local disposal regulations should be followed. Although the half-life of ^{32}P is as short as 14 days, isotope emission is only reduced by approx. 5% per day. The advantage of radio-labelled probes is that signal can be increased six times by prolonging the exposure from 1 day to 7 days, provided that the background is clean.

[b] If a very weak signal is expected, two intensifying screens may be used for one X-ray film. This further increases the sensitivity of the autoradiography. In this case, place the blot with the nucleic acids side up on an autoradiographic cassette. Lay a piece of film, sandwiched between two intensifying screens, on to the blot.

2.3.2 Detection of non-radioactive probes

Although several non-radioactive labels for nucleic acid detection have been described, biotin is probably the most commonly used. *Protocol 9* was developed for the detection of biotinylated probes (5) but can be applied with minor empirical modification to the detection of other reporter molecules. Theoretically, the non-radioactive method is as sensitive as that using radio-labelled probes. However, the former method is more technically demanding. This is because:

(a) The sensitivity of this method is highly dependent on the incorporation of the reporter molecule into the probe. Higher incorporation usually gives a stronger signal but decreases the hybridization efficiency of the probe. By contrast, lower incorporation interferes less with the hybridization reaction but reduces the sensitivity of detection. Therefore, optimal incorporation must be determined experimentally. In general, 30% substitution is adequate for both detection and hybridization.

(b) The detection reagents used in these systems may be quite labile. Partially degraded reagents can cause decreased sensitivity, high background, and troublesome non-specific staining, although this is not generally a problem with the high-quality reagents now available.

(c) Since the enzymatic reaction for signal detection only lasts for approx. 24 h, prolonged incubation in substrate solution does not increase the signal and usually results in high background. Therefore, unlike radioactive probes, the sensitivity of these systems is limited by the finite time of the enzymatic reaction.

Protocol 9. Detection of biotinylated probes

Materials

- 10 × STM buffer: 1 M Tris–HCl, pH 7.5, 1 M NaCl, 50 mM MgCl$_2$
- Blocking buffer
  - For nitrocellulose filters: 0.5% (v/v) Tween-20 in 1 × STM
  - For nylon filters: 0.5% (v/v) Tween-20, 1% non-fat dry milk [or 0.5% (w/v) casein] in 1 × STM
- Incubation buffer: 0.05% (v/v) Tween-20 in 1 × STM
- Streptavidin alkaline phosphatase conjugate (Dakopatts)
- Nitroblue tetrazolium (NBT): 75 mg/ml in 70% dimethylformamide (DMF)
- 5-bromo-4-chloro-3-indolyl phosphate (BCIP): 50 mg/ml in DMF
- Wash buffer: 0.5% (v/v) Tween-20 in 1 × STM
- Substrate buffer: 0.1 M Tris–HCl, pH 9.5, 0.1 M NaCl, 10 mM MgCl$_2$, 0.1 mM ZnCl$_2$

Method

1. After post-hybridization washing (see *Protocols 3* and *7*), rinse the blot in blocking buffer several times and block the filter with blocking buffer for 60 min at room temperature.

2. Incubate the filter in optimally diluted detection reagents for 10–30 min at 22–37°C depending on the reagents used.[a] For the biotin system, 10 min incubation in streptavidin alkaline phosphatase at room temperature is adequate. However, the suppliers instructions for the storage and use of these reagents should be followed.

3. Wash the filter in three changes of wash buffer for 5 min each then in two changes of substrate buffer for 5 min each.

4. Make up fresh substrate solution for the colorimetric detection of alkaline phosphatase by adding 44 µl of NBT solution and 33 µl of BCIP solution to 10 ml of substrate buffer and incubate the filter in this solution for 16–20 h in a plastic Petri dish in the dark.[b]

Protocol 9. *Continued*

5. After colour development, wash the filter extensively in water.

---

[a] The appropriate dilution of detection reagent should be determined empirically. For biotinyl-ated probes, dilutions of streptavidin alkaline phosphatase in incubation buffer over the range 1:250–1:10 000 would be appropriate.

[b] Alternatively, alkaline phosphatase activity can be demonstrated by chemiluminescence by placing the hybridized blot in a plastic bag, and adding substrate chemiluminescence solution (substrate buffer containing 0.25 mM chemiluminescent substrate, 4-methoxy-4-(3-phosphatephenyl) spiro[1,2-dioxetane-3,2′-adamantane] (Sigma)). Use 1 ml of substrate solution per 100 cm² of filter. Squeeze out all the air-bubbles carefully and seal the bag. The light emitted by the chemiluminescent reaction can be recorded on X-ray film (as for autoradiography; see *Protocol 8*) by exposure for 0.5–3.0 h depending on the abundance of the target sequences.

---

3. Preservation of hybridized blots

After probe detection, blots can be stored damp at 4°C. Alternatively, hybridized probes can be stripped from the blots which can then be reprobed if necessary or stored dry at 4°C. In general, hybridized probes can be removed by washing the blots in 0.025% SDS at 68°C for 3× 30 min. Stripping may not be necessary if radioactive blots have been stored for a long time before they are re-probed as the activity of the isotope after 5 half-lives is only $1/2^5$; that is, approx. 3% of its initial value.

For non-radioactive blots, although the detection reagent (alkaline phosphatase) degrades rapidly, the probe reporter molecules are very stable. Therefore, stripping of hybridized probe from the filter before re-probing is essential. However, re-probing is not recommended if colorimetric detection of non-radioactive probes is used, especially on nitrocellulose filters. This is because the end product of the enzymatic reaction, formazan, has to be dissolved using DMF which also dissolves the nitrocellulose. Although nylon filters are relatively resistant to DMF, the stripping of formazan and hybridized probes is quite tedious and time-consuming. Furthermore, the permanent record of the hybridization results is lost.

References

1. Hames, B. D. and Higgins, S. J. (ed.) (1985). *Nucleic acid hybridisation: a practical approach*. IRL Press, Oxford.
2. Southern, E. M. (1975). *J. Mol. Biol.*, **98**, 503.
3. Thomas, P. S. (1980). *Proc. Natl. Acad. Sci. USA*, **77**, 5201.
4. Seed, B. (1982). In *Genetic engineering: principles and methods* (ed. J. K. Setlow and A. Hollaender) Vol. 4, 91. Plenum Publishing, New York.
5. Wahl, G. M., Stern, M., and Stark, G. R. (1979). *Proc. Natl. Acad. Sci. USA*, **76**, 3683.

6. Chan, V. T. W., Fleming, K. A., and McGee, J. O'D. (1985). *Nucleic Acids Res.*, **12,** 8083.
7. McMasters, G. K. and Carmichael, G. G. (1977). *Proc. Natl. Acad. Sci. USA,* **74,** 4835.
8. Lehrach, H., Diamond, D., Wozney, J. M., and Boedtker, H. (1977). *Biochemistry,* **16,** 4743.

4

The polymerase chain reaction and the molecular genetic analysis of tissue biopsies

DARRYL K. SHIBATA

1. Introduction

The polymerase chain reaction (PCR) was described in 1985 by Mullis and co-workers at the Cetus corporation (1, 2). The subsequent inclusion of a thermally stable DNA polymerase (3) greatly facilitated its practical application and, since its introduction, the uses and users of the PCR have expanded exponentially in both research and clinical applications. This chapter will describe briefly the theory and methods of PCR in the molecular genetic analysis of tissue specimens. It will concentrate on the practical aspects and outline procedures applicable to routinely obtained tissue biopsy specimens.

1.1 PCR

PCR allows specific *in vitro* enzymatic replication of nucleic acids directed by synthetic oligonucleotide primers. The reaction is illustrated in *Figure 1*. The basic PCR cycle involves heat denaturation of the DNA, followed by primer annealing and then DNA synthesis. These three components can be achieved simply by altering the temperature of the reaction. The primers are orientated such that overlapping replication occurs, resulting in an exponential increase of nucleic acids. Several excellent handbooks (4, 5) discuss features and parameters of the reaction other than those which will be covered here. It is suggested that readers unfamiliar with PCR also survey these reviews.

The application of molecular techniques to the study of clinical specimens has been complicated by their lack of sensitivity and the requirement for prolonged assay times. Thus, radiolabelled nucleic acid hybridization probes generally require several thousand copies of target for detection. Therefore, many specimens containing clinically significant numbers of abnormal sequences (but in less than approximately 10000 copies) would be falsely negative. The PCR allows an assay to search specifically for a DNA sequence. If that sequence is present, the oligonucleotide primers (or 'amplimers'),

The polymerase chain reaction

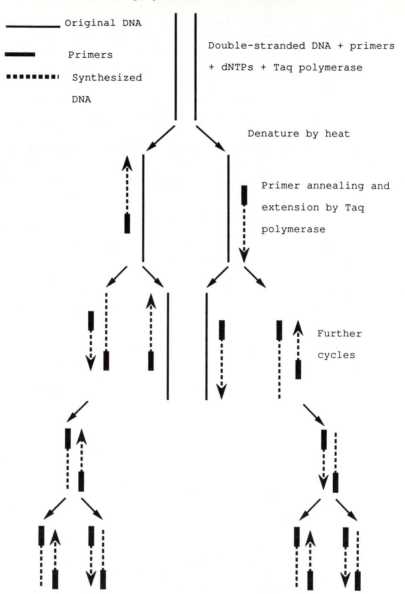

Figure 1. The principle of the polymerase chain reaction. The primers are chosen to produce specificity of amplification resulting in a product of defined size.

present in vast excess, anneal and allow *in vitro* DNA replication. The amplimers hybridize to opposite strands and therefore both the original target sequence and the synthesized PCR products serve as PCR substrates. The result is an exponential increase in the DNA sequence of interest (see *Figure 1*). The targets in many specimens falsely negative by conventional techniques are easily detected after PCR since as few as one target copy can be amplified to detectable amounts. The PCR can literally pick the 'needle in the haystack' as one abnormal target can be detected in a background of 100 000 to 1 000 000 normal cells. This sensitivity is far greater than other commonly used molecular techniques and rivals or surpasses that achieved by microbiological techniques.

In addition to enhanced sensitivity, the exponential increase in target sequences facilitates and simplifies their detection. Each temperature cycle typically requires 2 to 5 minutes. For most applications, the 25–50 cycles needed for the desired sensitivity require 2 to 4 hours. Therefore, assay times can be significantly reduced as the amplified products can usually be detected in less than one day. The large number of PCR products also facilitates non-isotopic detection methods. A comparison between the PCR and other molecular techniques is given in *Table 1*. The PCR thus possesses two important and critical advantages for diagnostic application: high sensitivity and short assay time.

Table 1. Comparison between nucleic acid hybridization techniques

	PCR	Southern blot	*In situ* hybridization
Sensitivity	1/100 000[a]	1/100	10 targets/cell
Specificity	High	High	High
Assay time	1 day	1 week	1 day
Cell location	No	No	Yes

[a] 1 copy in 100 000 cells

1.2 Amplification of RNA

RNA can also be amplified by the PCR after initial conversion to a cDNA copy by the enzyme reverse transcriptase. Since RNA is more easily degraded than DNA, its detection is more difficult, particularly from suboptimal clinical specimens. This chapter will concentrate on techniques which use DNA as the PCR substrate.

2. General considerations

Many variables can affect the PCR. The previously mentioned reviews (4, 5) provide information on the optimization of the PCR. Important factors include the number of cycles, time–temperature PCR profiles, and concentrations of the amplimers, magnesium, deoxynucleoside triphosphates, and enzyme. Standardized PCR reaction mixes are commercially available (for example, from Perkin–Elmer Cetus). The use of PCR has expanded such that PCR assays to detect a variety of target sequences have been published or are commercially available. It is anticipated that in the clinical setting, custom-designing of PCR assays will seldom be necessary or desirable. In this context, evaluation of the suitability of PCR tests for clinical application or their conversion to clinical assays will be presented.

2.1 Selection of primers

PCR primers or amplimers are generally 18–25 nucleotides long. They can be synthesized on automated DNA synthesizers which are present in most core facilities. The amount synthesized (0.2 to 1.0 μM) is usually sufficient for many hundreds to thousands of tests. Primers can also be obtained from many commercial sources (for example, DuPont, Perkin–Elmer Cetus). In many cases, restriction enzyme sites are incorporated at the 5′-ends of the primers to facilitate cloning of the PCR products. For most clinical applications, this is unnecessary.

The selection of primers is critical since they will define the sequence amplified and detected. This aspect provides enormous flexibility to the assay since, by simply varying the primers, many different pathogens or molecular genetic alterations can be detected by using essentially the same general procedures. Mastery of the technique for the detection of one target usually facilitates learning the detection of others.

PCR products have a distinct size defined by the sum of the lengths of the primers and the distance between their 3′-ends. In most instances, PCR products are 100–300 nucleotides long. Targets greater than 1000 base pairs can be amplified but generally with less success. (The *Taq* DNA polymerase replicates approximately 35–100 nucleotides per second.) In clinical practice, where tissue specimens may be less than optimal, DNA templates may be degraded (see below). Therefore, smaller target sequences are desirable since they can be detected in severely degraded tissues.

2.1.1 Conserved targets

Amplimers must hybridize to the desired target sequence to allow a successful PCR. If the desired target sequence is variable between individuals or microbiologic isolates, mismatches between these variations (or polymorphisms) and the amplimers or probe may prevent detection and result in a false-negative interpretation. Fortunately, most human genomic sequences are conserved between individuals and generally one defined set of amplimers homologous to the conserved targets ('generic' primers) will be adequate. An example of generic amplimers and a generic probe to such conserved sequences is given in *Figure 2* (Type 1).

```
     1              2                      3                    4

A    B    C    D    E    F              G    H    I         J    K         L

--------------------xxxxxxx----------zzzzzzzzzzzzzzz-------VNTRVNTRVNTR---


A    B    C    D    e    F              g    h    i         J.   K    L

--------------------yyyyyyy----------wwwwwwwwwwwwwww-------VNTRVNTR-------
```

Type	Amplimer	Probe	Examples
1.	Generic (A, C)	Generic (B)	Most human targets
2.	Generic (D, F)	Allele Specific (E, e)	Multi-allelic targets (sickle cell)
3.	Variable (G, I, g, i)	Type Specific (H, h)	Typing of viruses
4.	Generic (J, L)	Generic (K)	VNTR[a] amplification for identification

[a] Variable number tandem repeats.

Figure 2. General types of amplimers and probes.

2.1.2 Polymorphic targets

Primers directed against unexpected polymorphic targets may fail to hybridize and extend due to lack of homology. Mismatches in the 3' portion of the amplimer are most critical in this aspect. Studies by Kwok *et al.* (6) have demonstrated that some lack of homology in the terminal 3' primer nucleotide can be tolerated at the nucleoside triphosphate concentrations normally present during PCR. In particular, a primer with a terminal T will be extended regardless of the template mismatch. At lower nucleoside triphosphate concentrations, 3' template mismatches are not tolerated, and allele-specific amplifications based on single base-pair template–3' primer mismatches are possible (7).

Polymorphisms between individuals or isolates are inevitable. The only question is where these polymorphisms exist and in what frequency. Polymorphisms cannot be predicted and are revealed only after the analysis of many isolates. For human genomic targets, polymorphic loci are usually well-characterized and appropriate primers and probes can be designed to avoid them. In many instances, the polymorphism will be an important somatic or germline mutation and primers and probes can be specifically designed to detect these. Polymorphisms also allow discrimination between individuals (PCR 'fingerprinting' (8)).

For micro-organisms, unexpected polymorphisms between isolates can cause a false-negative PCR. The polymorphisms may reflect allelic differences or reflect inherent genomic instability such as is present in retroviruses. In practice, since there is generally a limited amount of sequence data available for most microbiologic organisms, the specificity of a primer and probe set should be determined empirically. An ideal set will detect all isolates of the desired species but not closely related but medically different species. For example, if a primer and probe combination is designed to be specific for one member of the herpes family, other herpes viruses should not be detected. Conversely, all isolates of the designated herpes species should be amplified and detected by the primers and probes. Because every isolate may not be conserved with respect to the target sequence, two different primers and probe probe sets against two different targets are often employed since it is unlikely that both targets will be polymorphic in the same isolate.

The strategy described by Manos *et al.* (ref. 9, and see Chapter 6) for the discrimination of human papillomaviruses (HPVs) takes advantage of the presence of both conserved and polymorphic regions by using 'generic' primers against conserved regions to amplify a large number of HPVs isolates and then type-specific probes against polymorphic sequences to distinguish between these isolates (Type 2 in *Figure 2*). Alternatively, different sets of primers and probes, both directed against polymorphic regions, may be designed to amplify specifically and distinguish between targets (Type 3 in *Figure 2*).

'Nested' primers can also be used to improve specificity and sensitivity. A second PCR is performed on a small portion of the PCR products from a first amplification, but using a set of primers which recognize sequences internal or 'nested' to the target sequences of the first PCR. The need to manipulate PCR products prior to a second amplification increases the probability of contamination and therefore nested primers are less appropriate for routine clinical use.

2.2 Number of cycles

The number of cycles helps to determine sensitivity. In general, higher numbers of cycles increases sensitivity. However, optimal sensitivity cannot be obtained by just increasing the number of cycles, and other parameters should also be varied to improve the amplification efficiency of each cycle. The prevalence of the desired target in the specimen will also determine cycle number. A single-copy gene expected to be present in every cell (such as a PCR assay for a genetic disease) would require a minimum of cycles (20–30) whereas a target expected to be present in small number of cells (such as a PCR assay for the detection of a virus or minimal residual tumor) would require a greater number of cycles (30–50).

2.3 Importance of fidelity

The DNA polymerase currently used in most PCR assays is isolated from *Thermus aquaticus*. It lacks 3′ to 5′ proofreading activity and has an error rate of approximately 1/10 000 bases. This level of fidelity, although less than many other polymerases, has allowed sufficiently accurate replication of target sequences that no practical problems have been encountered from mutations arising during amplification.

3. Preparation of tissue specimens for PCR

Since histological examination is adequate and diagnostic for most specimens, tissues are routinely formalin fixed and paraffin embedded for routine surgical pathology. The PCR can utilize such fixed tissues and therefore can be readily applied to the minority of cases which need additional analysis. This section will concentrate mainly on the application of PCR to the analysis of such fixed tissues.

3.1 Fresh tissues

DNA isolated from fresh tissues by conventional procedures (see Chapter 1) is optimal for the PCR. As the efficiency of amplification with purified DNA is usually greater than with DNA extracted from fixed tissue, caution should be taken when directly comparing the two substances.

3.2 Fixed tissues

3.2.1 Fixative

Ideally, tissue should be fixed immediately, although delayed fixation is not a major problem. Autopsy tissues with post-mortem intervals of up to seven days (and probably more) can be amplified. The DNA in tissues fixed in 10% buffered formalin and embedded in paraffin wax is an excellent PCR substrate (10, 11). Indeed, using careful extraction techniques, the DNA may be of sufficient quality with a high enough molecular weight for Southern blot analysis (refs. 12 and 13, and see Chapter 10). Paraffin-embedded tissues fixed in ethanol or acetone are also adequate PCR substrates. Somewhat poorer amplification is obtained with tissue fixed in Zamboni's and Clarke's fluids, paraformaldehyde, formalin–alcohol–acetic acid, and methacarn. Fixatives which should be avoided include Zenkers, Carnoy's, Bouin's, and 'B-5', as the PCR is usually not successful (see Greer *et al.* (14) for an excellent study of fixatives). Acidic solutions used for decalcification of bone usually degrade the template and prevent amplification.

3.2.2 Length of fixation

Prolonged fixation (greater than 3 days) in 10% buffered formalin should be avoided: the ideal fixation time is from one hour to 24 hours before placement in the tissue processor. It should be noted that additional fixation in 10% formalin usually occurs in most tissue processors.

Formaldehyde fixation results in the generation of Schiff bases with the DNA which are reversible in aqueous solutions. With prolonged fixation, other reactions occur which are not readily reversible. Once embedded in paraffin, the DNA is relatively stable. Unfortunately, slow, continued degradation of the DNA appears to occur and very old blocks (greater than 10 years old) may have reduced amplification efficiencies. However, DNA from most 40-year-old blocks can be analysed by PCR (15), and archaeological specimens have been used as PCR substrates.

3.2.3 Cutting of sections

The goal is to place a single slice of tissue in a sterile 1.5-ml microfuge tube without cross-contamination by other tissues. Usually, this can be accomplished by histotechnicians, although they may frown on squashing sections. A single 5–10-micron-thick section is placed in a sterile tube. The section can have an area from less than $1 \, mm^2$ to $1–3 \, cm^2$ and the procedure can be performed either with or without gloves. It is extremely difficult to cut sections with gloves and they are not routinely used in our laboratory. It appears adequate to clean the fingers by washing or with paper towels between sections. Some contamination of the tissue with the cells of the operator may occur, and this should be recognized during interpretation. Some operators to avoid may be those with dermatological diseases (particularly with parakaratosis), warts (if looking for papillomavirus), etc. A more dangerous type of contamination is carry-over from case to case. To avoid this, one should autoclave (for approximately 20 min) and carefully lay out and label the microfuge tubes before sectioning. The microtome should be kept relatively clean of wax and tissue fragments and the blade should be carefully wiped between blocks with paper towels. Generally, the blade does not need to be changed or wiped clean with xylene or acids. It appears that the normal facing of the block cleans the blade adequately between blocks. To check on technique, negative controls can be cut alternately with positive controls.

Paraffin blocks are not normally trimmed of excess paraffin. To analyze a selected portion of the tissue, it can be re-embedded, or the block may be scored with a scalpel such that, when cut, the area of choice can be separated from the undesired tissues. To verify the tissues obtained, an adjacent 'mirror' section should be stained with haematoxylin and eosin.

After cutting, sections are pushed into a 1.5-ml microfuge tube. They should be dry, as water will inhibit the removal of paraffin. For very small sections, it is better to place them directly into an empty PCR reaction tube (0.5-ml microfuge tube). The cut sections can be stored for months at room temperature.

Most tissue types are adequate PCR substrates. Some problems are encountered however with brain, spleen, and extremely bloody or necrotic tissues.

3.2.4 Extraction of DNA from tissue sections

Paraffin is removed from sections by conventional means as described in *Protocol 1*. Two basic procedures are used for DNA extraction depending on the size of the tissue. Procedures for small and large tissue fragments are described in *Protocols 2* and *3* and indications for their use summarized in *Table 2*. Another procedure which utilizes detergents has been described by Wright and Manos (17).

Table 2. Guide to nucleic acid extraction protocols

Area of section	Number of slices	Extraction protocol
<1 mm² (cell blocks of fine needle aspirates or very small biopsies)	1–3	*Protocol 2*
1–2 mm² (small biopsies)	1–3	*Protocol 2* or *3*
>2 mm² (most tissues)	1	*Protocol 3*

Protocol 1. Deparaffinization of sections

Sections can be deparaffinized with either xylene, octane, or Histoclear as follows:

1. Add approx. 1 ml of the preferred solvent to each 1.5-ml microfuge tube containing a tissue section, shake, and allow to sit for 2 min at room temperature. The paraffin should dissolve completely.
2. Centrifuge at 12 000 g for 5 min to pellet the tissue.
3. Decant the supernatant (the pellet may be difficult to see).
4. Add 1 ml of 95–100% ethanol and shake (the tissue should be slightly white in colour).
5. Centrifuge at 12 000 g for 5 min and decant the supernatant.
6. Repeat steps 4 and 5 (a total of two ethanol washes).
7. Desiccate the tissue (using, for example, a speed vac or desiccator jar) for 30 min or until dry.
8. Desiccated tissue may be stored at room temperature for several weeks.

Protocol 2. Extraction of DNA from small tissue fragments (see *Table 2*)

For very small specimens, the tissue should be amplified directly, after deparaffinization.[a]

1. Add 50 μl of sterile distilled water to the desiccated tissue from *Protocol 1*.

2. Place the closed microfuge tube in a boiling water bath for 7 min, and chill on ice.

3. Add 50 μl of amplification mix[b] and disrupt the tissue with the pipette tip to allow access of the reagents to the DNA. The residual tissue will remain visible in the microfuge tube.

4. Proceed with the amplification on a thermal cycler (see *Protocol 4*).

---

[a] For this technique, it is better to place the paraffin sections directly into 0.5-ml microfuge tubes, since transfer of the tissue fragments after removal of paraffin is difficult (see *Protocol 1*).
[b] See *Protocol 4* for composition.

Protocol 3. Extraction of DNA from large tissue fragments (see *Table 2*)

Materials

*   Extraction buffer: 100 mM Tris–HCl, 1 mM EDTA, pH 8.0.

*   Proteinase K solution: 20 mg/ml in sterile distilled water.

Method

1. Add 50 μl extraction buffer[a] to the desiccated tissue produced by *Protocol 1*.

2. Add 1 μl proteinase K solution.

3. Disrupt the tissue pellet with the pipette tip to allow access of reagents to the tissue and incubate at 37°C overnight or 50°C for 3 h.

4. Vortex for 30 sec and boil for 7 min to inactivate the proteinase K.

5. Spin at 12 000 g for 5 min to pellet the residual tissue fragments and use the supernatant for subsequent PCR.[b]

---

[a] The volume can be altered according to the size of the tissue section. The final concentration of proteinase K should be 400 μg/ml.
[b] DNA extracted in this way can be stored at either −20°C or 4°C for several months.

3.2.5 PCR

For very small tissue specimens, it is best to boil and amplify the entire section (see *Protocol 2*). For larger specimens, approximately 1/10th to 1/50th of the extraction solution should be used. In some cases, amplification is inhibited by the extracted DNA (or some other factor present in the extraction solution) and smaller amounts will be successful. The amount and degree of degradation of the extracted DNA can be determined by electrophoresis of 5–10 μl of the DNA solution obtained using *Protocol 3* on a 0.7% agarose gel stained with ethidium bromide (see Chapter 1).

The size of the target DNA sequence is critical in the analysis of fixed tissue. In general, smaller targets are more easily detectable than larger ones. Targets should ideally be less than 200 base-pairs (bp) in length and those less than 100 bp long will inevitably be successfully amplified.

PCR is performed by incubating the samples at three different temperatures corresponding to the three steps in a cycle of amplification—denaturation (90–95°C), annealing (40–60°C), and extension (72°C) (see *Figure 1*). The equipment used for thermal cycling may be as simple as water baths set at different temperatures for manual reactions or, increasingly commonly, automated microprocessor-controlled machines (available from, for example, Perkin–Elmer Cetus Instruments, Techne, and Koch-light).

A typical PCR protocol is given in *Protocol 4*. It is best to make a master mixture as described in step 1, distribute this to the reaction tubes and add the target DNA last. The reaction volume is not critical (accuracy to within approximately 10% is adequate) and adjustment for varying amounts of added specimen need not be made. The details of the amplification cycle given in step 4 are suggested as a starting point but the denaturation temperature of 94–95°C is applicable to all targets. Initial prolonged heating (5 min at 95°C) ensures complete denaturation of all sequences in the sample and subsequent denaturation in each cycle is performed for a shorter time (45–60 sec). Insufficient heating during the denaturation step is a common cause of failure in the PCR reaction. The annealing temperature determines the stringency of primer hybridization to target sequences and hence the specificity of amplification and depends on the length and G–C content of the primers. An approximation of the temperature can be derived using the formula:

$$4°C \times (G + C) + 2°C \times (A + T) - 3.$$

In practice, the temperature should be evaluated by empirical variation using positive and negative controls: a suitable range to try would be 37–65°C.

The extension temperature should be 72°C as this gives optimal *Taq* polymerase activity. The required extension time is determined by the length of the amplified product sequences and a rough rule of thumb is that 1 minute extension is required for every 1000 bp of target DNA. Therefore, for archival specimens, in which targets 100–200 bp long are analysed, extension times of 45–60 sec are usually adequate.

95

Protocol 4. The polymerase chain reaction[a]

Materials

- 10 × buffer: 100 mM Tris–HCl, pH 8.3, 500 mM KCl, 25 mM MgCl$_2$
- 10 × dNTPs: 2 mM each of dATP, dCTP, dTTP, and dGTP
- Primers: 20–100 pmol of each primer per reaction
- *Taq* polymerase
- Extracted DNA (from *Protocol 2* or *3*)
- Mineral oil

Methods

1. Mix the following:
   - 10 × buffer   10 μl
   - 10 × dNTPs   10 μl
   - Primers   to give a final concentration of 0.2–1 μM
   - *Taq* polymerase   0.4 μl (2 Units)[b]
2. Add the sample DNA and make up to 100 μl with sterile distilled water.[c]
3. Overlay with 1–2 drops of mineral oil.
4. Place the tubes in a thermal cycler programmed appropriately. A typical cycle would be: initial denaturation for 5 min at 94 °C followed by 20–50 cycles of; denature at 95 °C for 45 sec, anneal at 50 °C for 45 sec and extend at 72 °C for 60 sec (see text).

---

[a] The reaction should be optimized for each combination of primers and target DNA by manipulation of magnesium, dNTP, and primer concentrations using appropriate positive and negative controls. The concentrations given in this *Protocol* are designed as an appropriate starting point.

[b] The amount of *Taq* enzyme can be increased (to 4 Units or more) if the reaction appears to be inhibited by substrate DNA.

[c] A 50 μl reaction volume can be used by scaling down the amounts given in order to save reagents.

3.3 Cerebrospinal fluid (CSF)

For cerebrospinal fluid, the small volume and paucity of cells prevent the utilization of most purification methods or the analysis by other molecular techniques. In my experience and others (18, 19), non-bloody CSF can be utilized directly from the patient after boiling. Up to 50 μl of CSF can be added to a 100 μl PCR final volume. A typical procedure is described in *Protocol 5*.

Protocol 5. PCR of cells present in CSF[a]

1. Determine the number of white blood cells present in the sample (for example 10 cell/μl)[b]

2. Take 1 ml and dispense as follows into sterile 0.5 ml PCR tubes:

Amount of CSF (μl)	Water (μl)	Cells (for concentration of 10/μl)
10	40	100
50	0	500
200[c]	0	2000
750[c]	0	7500

3. Boil for 5–10 min to disrupt the cells.

4. Add PCR mix and amplify according to *Protocol 4*.

[a] The PCR can also measure extracellular organisms (concentrated by ultracentrifugation) in CSF (ref. 20, and see Chapter 8).

[b] The maximum number of crudely prepared cells which should be amplified is approximately 100 000.

[c] For these dilutions, spin down the cells in a clinical centrifuge and remove the supernatant to leave a residual volume of 50 μl, which is then used for PCR.

3.4 Tissues frozen in optimal cutting temperature compound (OCT)

Another common clinical problem is tissue frozen in freezing compounds (such as OCT used for frozen sections). These compounds aid in sectioning of the tissue. A single 10-micron frozen section may be rapidly and easily processed for PCR, thus saving the bulk of the remaining tissue for further studies (see *Protocol 6*). The PCR performed on this DNA is usually more efficient than that using DNA extracted from fixed tissues.

Protocol 6. Preparation of single frozen tissue slices

1. Place a single 10 μm cryostat frozen section in a sterile tube.

2. Add a small volume of 0.9% (w/v) NaCl, vortex briefly, centrifuge at 12 000 g for 5 min to pellet the tissue and decant twice to remove the OTC.

3. Fix the tissue with 95% ethanol for 5–10 min.

4. Centrifuge at 12 000 g for 5 min and then desiccate.

5. Proceed with the DNA extraction as for paraffin-embedded tissues (see *Protocol 3*) and amplify the target DNA (see *Protocol 4*).

3.5 PCR of archival slides

DNA can also be amplified from stained archival slides or Papanicolau-stained smears. The tissues are scraped off the slides and then processed in a manner similar to deciccated deparaffinized tissue (see *Protocol 4*). The generation of numerous tissue fragments makes cross-contamination between specimens more likely and extreme care should be exercised.

3.6 *In situ* PCR

A promising technology is the ability to perform PCR with subsequent *in situ* hybridization. This would allow both high sensitivity and identification of the target containing cell. Initial progress has been reported (22).

4. Analysis of PCR products

There are many techniques for the analysis of PCR products. Three simple techniques, namely gel electrophoresis, dot blotting and Southern blotting will be outlined (see also Chapters 1, 3, and 6). The majority of PCR products can be analysed by these techniques but absolute identity can only be established by nucleic acid sequencing (see Chapter 5).

4.1 Preparation of the PCR product

Before analysis, the overlying mineral oil must be removed. This is achieved by adding one or two drops of chloroform to the PCR tubes, shaking briefly and spinning for 1 min at $12\,000\,g$ in a microfuge to separate the top aqueous phase which contains the PCR product. The product can be stored at $4\,°C$ with the chloroform for several weeks. In general, 1/10th to 1/5th of the reaction volume is subsequently analysed.

4.2 Gel electrophoresis

Agarose gel electrophoresis is simple, provides information on both the presence of a PCR product and its size and does not involve the use of isotopes. The general principles of agarose gel electrophoresis are described in Chapter 1 but points specific to the analysis of PCR products are:

(a) Analyse 10–20 µl of the PCR product together with a size standard of fragment size 50–1000 bp on a 2% agarose gel.

(b) High voltages (for example 10–15 V/cm) should be used since small PCR products may be more difficult to detect after an overnight run due to diffusion. Better resolution of the products can be achieved by using polyacrylamide gels (see Chapter 6) or high percentage (3–4%) NuSieve (FMC BioProducts, Rockland, USA) agarose gels. However, for quick analysis, a 2% agarose gel is relatively inexpensive and easy to use.

Lower yields and multiple non-specific PCR products are commonly encountered when amplifying fixed tissues by comparison with purified DNA
templates. The visualized result on an agarose gel is often a smear and/or
multiple bands and simple gel electrophoresis therefore lacks sensitivity and
specificity. A hybridization step, as described in Sections 4.3 and 4.4 usually
improves the analysis.

4.3 Analysis of PCR products by Southern blotting

This technique allows the specific identification of bands present after gel
electrophoresis as PCR products and improves both the sensitivity and
specificity of products analysis. Both non-isotopic and isotopic hybridization
probes can be used and the procedure is outlined in *Protocol 7*.

Protocol 7. Southern blot analysis of PCR products[a]

Materials
- Denaturing buffer: 1.5 M NaOH, 0.5 M NaCl
- Neutralizing buffer: 1.5 M Tris–HCl, 0.5 M NaCl, pH 6.5
- 20 × SSPE: 3.6 M NaCl, 0.2 M sodium phosphate, 0.02 M EDTA, pH 7.4
- 10% (v/v) SDS

Method
1. Float the gel in denaturing buffer to denature the DNA fragments for
   30 min at room temperature with gentle shaking, then in neutralizing
   buffer for 30 min.
2. Meanwhile, soak a nylon transfer membrane in 6 × SSPE for at least 5 min. In
   our laboratory, we use Genetrans 45 (Plasco, Wolburn, Massachusetts,
   USA).
3. Arrange the gel, nylon filter, and filter papers as shown in *Figure 1*,
   Chapter 3, and transfer the fragments overnight at room temperature
   using 20 × SSPE as the transfer buffer.
4. Remove the filter papers and mark the positions of the gel wells on the
   nylon membrane. Wash the nylon membrane in 20 × SSPE for 1 min to
   remove agarose fragments.
5. Place the nylon membrane, DNA side down, on a 254 nm (short wave)
   UV light box and irradiate for 5 min to crosslink the DNA to the filter.
   Alternatively, use a UV crosslinker (Stratagene, La Jolla, California,
   USA).
6. Place the membrane in a sealable plastic bag,[b] pre-hybridize at 42 °C for
   10 min then hybridize at 42 °C for 30–60 min (see Chapter 3) with the

Protocol 7. *Continued*

appropriately labelled probe (see Chapter 2). The most commonly used probes are oligomer sequences labelled at the 5'-end with T4 polynucleotide kinase.

7. Wash the membrane at room temperature with 2–3 changes of 2 × SSPE, 0.1% SDS for 5 min each. Higher stringency washes can be performed at this stage but this requirement is determined empirically.[c]

8. Detect the presence of hybridized probe by either autoradiography or non-isotopic detection methods (see Chapter 3).

[a] For further details, see Chapter 3.
[b] Membranes prepared in this way can be stored at room temperature for several weeks.
[c] Stringency can be elevated by increasing the concentration of formamide, decreasing the salt concentration or elevating the washing temperature. Commonly used combinations are: 0.1 × SSPE at 65°C; 50% formamide, 2 × SSPE at 37–42°C.

4.4 Analysis of PCR products by dot blotting

The dot blot provides a simple yes/no answer, It requires a hybridization probe and is useful if many samples are analysed. An example procedure is given in *Protocol 8* and further details can be found in Chapter 3.

Protocol 8. Dot blot analysis of PCR products

Materials
- Denaturation solution: 0.4 M NaOH, 1 mM EDTA
- 20 × SSPE (see *Protocol 7*)
- Nylon membrane
- Dot blot manifold[a]

Method
1. Wet the nylon membrane in distilled water for at least 5 min.
2. Take 10–20 μl of PCR product and add it to 300 μl denaturation solution (denaturation is essentially instantaneous).
3. Spot each sample into wells. Each well should take about 1 min to empty.[a]
4. Wash each well with 20 × SSPE, three times (5 min each).
5. Remove the filter and rinse with 20 × SSPE.
6. Proceed to step 5 of *Protocol 7*.

[a] Alternatively, samples can be applied directly to the membrane by hand (see Chapter 3 for details).

5. Interpretation of results

The results of a PCR may be interpreted as positive or negative, implying respectively that the target of interest was present or absent. However, false-negatives can occur secondary to inhibition of amplification, lack of sensitivity of the assay, or unexpected polymorphisms between the primers or probes and the target. False-positives occur secondary to contamination, or un-expected homology between the primers and probes to a related but different target. The problem of primer and probe cross-reactivity or unexpected polymorphism have been covered in Section 2.1.2.

5.1 Selection of controls

This is a critical aspect for meaningful interpretation. Well-characterized positive and negative controls should be utilized. In many instances, DNA from cell lines either possessing or lacking the desired target can be used. At least three controls should be present in every assay:

- Positive control
- Negative control
- Water blank

The water blank consists of everything but added specimen and is a sensitive indicator of contamination. If possible, the positive controls should consist of a high positive comprising many targets and a low positive containing only a small number of targets, so that the sensitivity and efficiency of the PCR can be monitored.

In some cases, no amplification occurs. This can lead to the problem of interpretation of potential false-negative results. In order to prevent this, each sample should always be tested by amplification of a genomic target, such as the low density lipoprotein receptor or β-globin gene (see Chapter 6 for practical details). This target should be slightly larger than the target under investigation to ensure that the sample DNA is not too degraded. The PCR should amplify every human sample for a genomic sequence. Lack of amplification of a genomic sequence therefore implies that the enzyme is inhibited or that the DNA template of the sample is either too degraded or present in too small an amount. The sample should therefore be re-amplified with both greater and lesser amounts of substrate DNA or more *Taq* enzyme. In some cases, the reaction will fail despite these manipulations. In many of these cases, fixative other than 10% formalin has been used or the tissue is necrotic and there is no solution to the problem other than re-biopsy.

The genomic primers may be included in the same PCR as the target primers and amplification carried out simultaneously as outlined in *Protocol 4* (and see Chapter 6). This multiplex PCR may have reduced sensitivity compared with a single set of primers, but the method ensures that false-negatives will not occur. A typical analytical scheme and interpretation of results is given in *Table 3*.

Table 3. Controlling for contamination in PCR

Tube	Sample	Target	Genomic	Interpretation
1	Tissue 1	+	+	Tissue positive
2	Tissue 2	−	+	Tissue negative
3	Tissue 3	−	−	Not informative, repeat
4[a]	Tissue 4	+	−	Contamination?, repeat
5[b]	+ Control	+	+	Expected
6[c]	− Control	−	+	Expected
7[d]	Blank	−	−	Expected

[a] Occasionally, this occurs. If contamination can be ruled out, then the tissue is probably positive for the target sequence. This pattern may occur if the target sequence is shorter or more abundant than the genomic sequence.
[b] Positive control: usually DNA isolated from a positive cell line.
[c] Negative control: usually DNA isolated from a negative cell line.
[d] Blank: everything but added DNA. This ensures that the buffers, nucleotides, primers, etc., are not contaminated.

5.2 Sensitivity

The exact sensitivity of a PCR assay on formalin fixed, paraffin embedded tissue is difficult to determine. Unfortunately, well-characterized, naturally occurring tissues with known amounts of target do not exist. It is unlikely that the sensitivity observed with purified DNA template under the same reaction conditions will match the sensitivity obtained from fixed tissues.

The availability of well-characterized cell lines allows the artificial construction of fixed tissues of known target content. Mixtures of cell lines with and without the target are fixed in formalin and paraffin embedded as a cell block. This block is sectioned and processed in the same way as biopsy specimens (see *Protocols 1–4*). PCR is performed and the sensitivity can be determined (see *Figure 3*). Although this is not a perfect sensitivity control (see below), it provides a guide for interpretation.

5.3 Quantitative PCR

Logically, the amount of PCR product should reflect the number of target sequences in the original specimen. Although this is generally true, often there are exceptions. These exceptions are particularly common when using paraffin-embedded tissue as a substrate as the number of variables which can effect the efficiency of amplification are numerous. They include the type of tissue, fixative, length of fixation, interval before fixation, age of the paraffin block, size of the section, and the amount of extracted DNA added to the

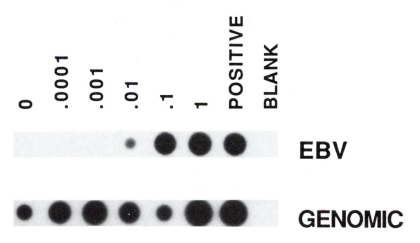

Figure 3. Sensitivity of the PCR on formalin-fixed, paraffin-embedded tissues. Mixtures of Raji cells (approximately 50 copies EBV/cell) and Molt 3 cells (EBV-negative) were pelleted in a clinical centrifuge and then fixed in 10% buffered formalin for one hour. The pellets were placed and then processed into paraffin embedded cell blocks. One to three 10 micron thick slices representing approximately 1000 cells were cut and processed by *Protocols 1* and *2*. PCR with both genomic and EBV specific primers was performed directly on the cell mixtures. A dot blot of the PCR products was performed in duplicate and hybridized against an EBV or a genomic probe. Under these conditions, the genomic signal could be detected in all mixtures (as expected) but the EBV signal could be detected from the 0.01 mixture but not from the 0.001 or 0.0001 mixtures. Therefore, under these PCR conditions, an average of approximately ten EBV infected cells per 1000 cells (the number of cells added to the PCR) could be detected. Better sensitivity was seen when the PCR conditions were changed and a greater number of cells were amplified.

PCR. All of these factors (and perhaps more) conspire to confound a proportional linkage between the final amplification signal and the original amount of target.

Quantitative PCR may also be difficult for many other clinical specimens which are received under less than ideal circumstances. Absolute quantitative analysis under these conditions would be extremely difficult. However, a simple method to determine the relative number of targets present in these suboptimal specimens is the PCR of serial dilutions of the specimen (see *Figure 4*). The specimen is diluted into water in a series of tenfold dilutions and then each dilution is amplified. The PCR is typically optimized for maximal sensitivity and/or 50 cycles are performed. This dilution analysis provides a minimum estimate of the number of targets present in the sample as theoretically no signal should be generated if the targets are sufficiently diluted. Since a genomic sequence is also amplified, relative comparison between the number of cells and the number of targets is possible. An example of this dilution analysis is presented in *Figure 5*. If a specimen contains a known number of cells (such as CSF), interpretation is as shown in *Table 4*.

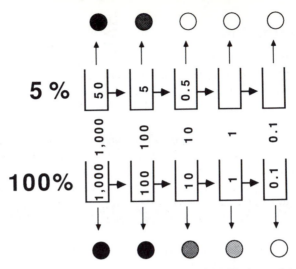

Figure 4. Estimation of the amount of target. Serial tenfold dilutions of the specimen are made. Typically, 1 μl of the specimen is diluted into 10 μl PCR tubes and then diluted further. PCR master mix is added to the 10 μl serial dilutions and the dilutions amplified. Specimens with lesser numbers of targets will become negative at lesser dilutions.

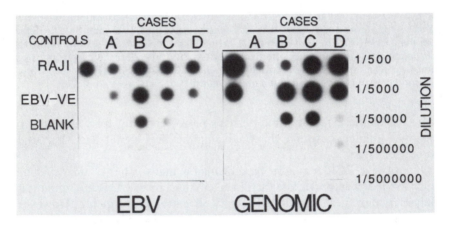

Figure 5. Relative amounts of target determined by dilution analysis. The DNA extracted from Cases A–D (*Protocol 3*) were serially diluted (see *Figure 4*) and then amplified with both genomic and EBV primers. The dilutions refer to the amount of the original section with 1/500 representing 0.1 μl of the original 50 μl DNA extraction solution from *Protocol 3*. The PCR products were dot blotted (*Protocol 8*) in duplicate and then hybridized against EBV or genomic labelled probes. The relative amounts of EBV DNA can be categorized into three groups. If the dilutions of the EBV and genomic targets are similar, they are classified as 3+ (specimens A, B, and D). If the genomic sequences are positive at greater dilutions than EBV, they are classified as 2+ (specimen C). If only the undiluted specimen was EBV positive, it would be called 1+. This type of classification allows relative comparisons between fixed tissue specimens which inevitably differ in cellularity, size, and fixation.

6. Contamination

There have been several excellent articles on laboratory practices designed to prevent contamination (24, 25) which should be read. Of particular importance is the segregation of reagents into small aliquots, since de-contamination procedures involve the discarding of suspected reagents. Contamination occurs when target sequences not present in a specimen are accidentally introduced prior to amplification. This may occur at any stage from the collection of the specimen from the patient, specimen processing, to the set-up of the reaction. Because of the exponential amplification provided by the PCR, minute quantities of contaminating DNA (such as one contaminating target) can result in a false-positive PCR. Therefore great care, analogous to microbiological techniques, must be exercised.

6.1 Severity and recognition of the problem

Contamination which results in false-positive results is a major drawback to the clinical application of the PCR. Two general types of contamination occurs: sporadic and generalized. Generalized contamination of one or more of the PCR reagents can be recognized when most specimens, and the negative controls, are positive. This type of contamination, although undesirable, should not result in false-positives since the controls will indicate its presence. Sporadic contamination is more insidious as, in this instance, individual specimens or reaction tubes become contaminated. This type of contamination is difficult to recognize since the false-positives will appear at random.

Table 4. Calculation of sensitivity by dilution analysis

Number of cells amplified	Target	Genomic	Interpretation
100 000	+	+	≥ 1 target/100 000 cells
10 000	+	+	≥ 1 target/10 000 cells
1 000	+	+	≥ 1 target/1000 cells
100	+	+	≥ 1 target/100 cells
10	+	+	≥ 1 target/10 cells
1	+	+	target in virtually every cell
0	−	−	Expected

The exact number of cells in the last dilutions is uncertain due to the Poisson distribution and pipetting errors. The genomic target should be amplified and detected to approximately the one cell level. Under these conditions, if 1% of the cells contained the target, the last dilution expected to be positive would contain 100 cells.

6.2 Laboratory practice

Prevention and recognition of sporadic false-positives depends on good laboratory set-up and techniques. Recognition may be aided by analysing each sample in duplicate or triplicate. All replicates should be positive if the sample is truly positive. In addition, PCR amplification of different targets should be performed if possible since it is less likely that contamination will occur with both targets. Finally, a completely new independent specimen can be obtained from the patient and re-analysed. Obviously, the requirement for a second specimen to verify the initial result is time-consuming and negates some of the time advantages of PCR. Good laboratory practice minimizes this need and the importance of a result may dictate the need for multiple verifications.

Another method for the recognition of sporadic contamination is to amplify polymorphic targets (see Section 2.1.2). VNTR, restriction site, and sequence polymorphisms can be utilized (26, 27). In a given PCR run, if an unexpected number of specimens produce the same polymorphism, contamination may be suspected.

There are generally two types of laboratory contamination: cross-contamination between specimens, which may be avoided by good laboratory practice and contamination by PCR products. The latter appears to be a greater problem since the number of PCR products, each of which can serve as a source of contamination, in a single PCR tube can be enormous. For instance, a single positive 100 µl PCR tube may contain 10^9 copies of target. If this tube was accidentally spilled into one hundred litres of water, 1 µl of this 'contaminated water' would contain an easily amplifiable 10 copies of potential contaminants. This sort of analysis illustrates how difficult it is to 'decontaminate' product spills and how easily 'everything' in a laboratory will appear to be positive for a given target. Awareness of the possible promiscuity of every reagent or device is of paramount importance.

It is of particular importance that the laboratory is set up inherently to prevent contamination. Since PCR product contamination is the most likely source of false-positives, the laboratory should be designed to isolate pre-PCR from post-PCR procedures physically. There should be a uni-directional flow of the procedure and products. The critical junction is where the PCR tubes are opened after amplification. This whole area should be considered contaminated and ideally nothing: samples, tubes, pipettes, centrifuges, etc., should be returned to the pre-PCR areas. It is often convenient to have the two areas present in separate buildings to emphasize this unidirectional flow. With such a laboratory, contamination by PCR products is inherently impossible and good laboratory techniques are usually sufficient to minimize cross-contamination between specimens.

Although gloves are often recommended, they are not routinely employed in our pre-PCR procedures. Most gloves are ill-fitting and hinder fine and

106

careful manipulation. They also provide a false sense of security unless changed routinely. Laminar flow hoods are desirable but not essential. Exposure of DNA to short-wave (254–300 nm) UV light appears to inhibit its ability to serve as a PCR template (28). Therefore, chronic exposure of the pre-PCR area to short-wave UV light out of working hours may reduce the number of stray targets.

Workers can be a major source of contamination. The rule of unidirectional flow also applies to them and it is risky to allow an individual to set up a PCR or process specimens after manipulating PCR products. Ideally, one individual should be in the pre-PCR area and another in the post-PCR area. No special precautions are necessary in the post-PCR area.

Positive controls are also a potential source of contamination since relatively large numbers of targets (plasmids, cells, etc.) can be produced through biological amplification. If possible, positive controls should be prepared outside the pre-PCR area with only small numbers of control targets utilized in the PCR set-up.

6.3 If contamination becomes a problem

Primarily every effort should be made to prevent contamination. In the event that contamination is detected, its source must be eliminated. The pattern of contamination will provide a clue to its source. Attempts to pin-point the exact reagent contaminated are often futile, time-consuming, and can spread the contamination. Therefore, all aliquots used to set up the contaminated PCR should be empirically discarded. If contamination persists, mechanical devices such as the desiccator, pipettes, or microtomes should be suspected and possibly changed or decontaminated with short-wave UV light or extensive cleaning. Contaminated areas or equipment may be identified by 'wipe' tests of the pre-PCR area (29). Laboratory procedures should be carefully reviewed.

If contamination persists, two options exist. The first is to change the target. Primer combinations directed at another target, or primers which anneal outside the original primer pair (reverse 'nesting') will negate PCR product contamination. A second option is to stop working for several months. Paradoxically, extensive efforts to decontaminate often lead to further contamination ['PCR Madness' or a 'PCR Sorcerer's Apprentice' (30)]. Often by stopping for 2–3 months, contamination subsides spontaneously (probably secondary to the natural degradation of nucleic acids). Of course, work can continue on other targets not affected by contamination.

7. Some applications in the clinical laboratory

The applications of the PCR to the clinical laboratory are numerous. They include the detection of viruses (see Chapters 6 and 7), mycobacteria,

protozoa, and bacteria (see Chapter 8), germline and acquired genetic alterations (see Chapter 10), and forensic identification (see Chapter 9). The
subject was recently reviewed by Eisenstein (31).

The routine use of the PCR in most clinical laboratories must await a
sufficient level of automation. Although the PCR itself is automated, the pre-
PCR and post-PCR steps require relatively extensive manipulations. These
manipulations are amenable to automation and most likely completely automated units similar to most modern clinical laboratory analysers will become
commercially available.

An intermediate step are kits such as the Amplitype kit (Cetus) for HLA
DQ alpha typing which minimize the number of manipulations. With this kit,
only $MgCl_2$ and the DNA specimen are added to pre-made PCR tubes. The
primers are biotin-labelled and the PCR products are detected non-
isotopically after hybridization to immobilized probes (8).

PCR tests can currently be performed by dedicated clinical laboratories.
In the context of routine surgical pathology, since formalin-fixed, paraffin
embedded tissues can be analysed, numerous specific molecular genetic
questions can be answered as they arise without the necessity of storing
frozen tissue. After histologic examination which is diagnostic and sufficient
for most specimens, a minority of specimens may be analysed by PCR
to answer specific questions. These specific questions may become more
common as the current clinical trend is to procure very small specimens
(fine needle aspirations, endoscopic biopsies, etc.) which can render histological diagnosis difficult. Some of the questions and possible approaches
include:

(a) Is the lesion reactive, or are there genetic alterations characteristic of
neoplasia present?

This can be one of the most difficult questions for a surgical pathologist.
Often the difficulty is secondary to suboptimal specimen handling (crush or
drying artefact) or the small specimen size. The PCR can analyse such
specimens easily and detect molecular genetic alterations which are associated with neoplasia. There is a steadily growing number of such alterations
and some of them are presented in *Table 5*. Unfortunately, there has been
little clinical experience with the PCR technique in diagnostic pathology, and
in most instances the molecular genetic alterations are not specific to neoplastic tissue, occurring also in normal and premalignant tissue. In addition,
the alterations are rarely tumour-specific. The information obtained by PCR
analysis, however, can be used as an adjunct to the clinical and histological
data and help render a more accurate diagnosis.

(b) Where is the primary lesion?

In many instances, a metastatic lesion is sampled. A diagnosis of malignancy
can be made but the primary site often defies histological examination. The
metastatic lesion can be examined by PCR for molecular genetic alterations

Darryl K. Shibata

Table 5. Brief list of some genetic alterations detectable by PCR.

Genetic alteration	Associated tumour types
K-*ras* point mutation	Adenocarcinoma • lung (cigarette smokers) • pancreas • colon
N-*ras* point mutation	Acute myeloid leukaemia Myelodysplastic syndromes Some acute lymphoid leukaemias Myeloma
t(14;18)	Follicular lymphoma
t(9;21)	Chronic myeloid leukaemia
t(8;14)	Burkitt's lymphoma
Viral infection	
Hepatitis B	Hepatocellular carcinoma
Human papillomavirus	Cervical warts, dysplasia, carcinoma Oropharyngeal neoplasms
HTLV-1	Adult T cell leukaemia/lymphoma
Epstein–Barr virus	Burkitt's lymphoma, nasopharyngeal carcinoma, Hodgkin's and non-Hodgkin's lymphoma

largely specific for certain primary sites. For example, the detection of human papillomavirus in a lymph node would indicate a likely oropharyngeal or an anogenital primary whereas detection of EBV in a carcinoma would indicate a likely nasopharyngeal primary.

(c) Is there minimal residual disease?

Morphological identification of rare tumour cells is difficult. For instance, less than one leukemic cell among 20 normal cells can rarely be identified microscopically in bone marrow biopsies. In contrast, the PCR is several logs more sensitive. Other specimens in which small numbers of tumour cells may be present are lymph nodes removed for staging, peripheral blood, CSF, urine, sputum, and various effusions.

PCR is used to identify targets which were present in the primary lesion. Detection of these tumour-associated targets (see *Table 5*) implies the existence of minimal residual disease. More specific detection of minimal residual disease is obtained by amplification of tumour specific targets. Rearrangements of immunoglobulin genes (32), T-cell receptor genes (33), or chromosomal translocations (34, 35) result in the generation of sequences which can be used to identify the neoplastic clone uniquely. The clinical significance of very small numbers of cells containing the tumour-associated marker is

109

currently unclear. The PCR provides objective measurement of low tumour burden and may allow rational therapy of subclinical disease.

(d) Does the tissue belong to the patient?

Occasionally, mis-labelling of specimens may be suspected. Using PCR against the HLA DQ alpha (Amplitype kit, Cetus) or any other polymorphic region, a PCR fingerprint can be obtained from virtually any clinical specimen or fixed tissue (36) and compared with the individual in question.

References

1. Mullis, K. B. and Faloona, F. A. (1987). In *Methods in enzymology,* Vol. 155 (ed. R. Wu), pp. 335–50. Academic Press, London.
2. Saiki, R. K., Scharf, S., Faloona, F., Mullis, K. B., Horn, G. T., Erlich, H. A., and Arnheim, N. (1985). *Science,* **230,** 1350.
3. Saiki, R. K., Gelfand, D. H., Stoffel, S., Scharf, S. J., Higuchi, R., Horn, G. T., Mullis, K. B., and Erlich, H. A. (1988). *Science,* **239,** 487.
4. Innis, M. A., Gelfand, D. H., Sninsky, J. J., and White, T. J. (ed.) (1990). *PCR protocols, a guide to methods and applications.* Academic Press, San Diego, California.
5. Erlich, H. A. (ed.) (1989). *PCR technology.* Stockton Press, New York.
6. Kwok, S., Kellogg, D. E., McKinney, N., Spasic, D., Goda, L., Levenson, C., and Sninsky, J. J. (1990). *Nucleic Acids Res.,* **18,** 999.
7. Ehlen, T. and Dubeau, L. (1989). *Biochem. Biophys. Res. Commun.,* **160,** 441.
8. Saiki, R. K., Walsh, P. S., Levenson, C. H., and Erlich, H. A. (1989). *Proc. Natl. Acad. Sci. USA,* **86,** 6230.
9. Manos, M. M., Ting, Y., Wright, D. K., Lewis, A. J., Broker, T. R., and Wolinsky, S. M. (1989). *Molecular Diagnostics of Human Cancer*, 7, 209–14. Cold Spring Harbor Laboratory, Cold Spring Harbor, NY.
10. Impraim, C. C., Saiki, R. K., Erlich, H. A., and Teplitz, R. L. (1987). *Biochem. Biophys. Res. Commun.,* **142,** 710.
11. Shibata, D., Arnheim, N., and Martin, W. J. (1988). *J. Exp. Med.,* **167,** 225.
12. Goelz, S. E., Hamilton, S. R., and Volgelstein, B. (1985). *Biochem. Biophys. Res. Commun.,* **130,** 118.
13. Dubeau, L., Chandler, L. A., Gralow, J. R., Nichols, P. W., and Jones, P. A. (1986). *Cancer Res.,* **46,** 2964.
14. Greer, C. E., Peterson, S. L., Kiviat, N. B., and Manos, M. M. (1991). *Am. J. Clin. Pathol.,* **95,** 117.
15. Shibata, D., Martin, W. J., and Arnheim, N. (1988). *Cancer Res.,* **48,** 4564.
16. Shibata, D., Brynes, R. K., Nathwani, B. N., Kwok, S., Sninsky, J., and Arnheim, N. (1989). *Am. J. Pathol.,* **135,** 697.
17. Wright, D. K. and Manos, M. M. (1990). In *PCR protocols: a guide to methods and applications* (ed. M. A. Innis, D. H. Glefand, J. J. Sninsky, and T. J. White), pp. 153–8. Academic Press, San Diego, California.
18. Shaunak, S., Albright, R. E., Klotman, M. E., Henry, S. C., Bartlett, J. A., and Hamilton, J. D. (1990). *J. Infect. Dis.,* **161,** 1068.

19. Shibata, D., Nichols, P., Sherrod, A., Rabinowitz, A., Bernstein-Singer, M., and Hu, E. (1990). *Mod. Pathol.*, **3**, 71.
20. Rowley, A. H., Whitley, R. J., Lakeman, F. D., and Wolinsky, S. M. (1990). *Lancet*, **335**, 440.
21. Jackson, D. P., Bell, S., Payne, J., Lewis, F. A., Sutton, J., Taylor, G. R., and Quirke, P. (1989). *Nucleic Acids Res.*, **17**, 10134.
22. Haase, A. T., Retzel, E. F., and Staskus, K. A. (1990). *Proc. Natl. Acad. Sci. USA*, **87**, 4971.
23. Sambrook, J., Fritsch, E. F., and Maniatis, T. (ed.) (1989). *Molecular cloning*. Cold Spring Harbor Laboratory Press, Cold Spring Harbor, NY.
24. Kwok, S. (1990). In *PCR protocols* (ed. M. A. Innis, D. H. Gelfand, J. J. Sninsky, and T. J. White), pp. 142–5. Academic Press, San Diego, California.
25. Higuchi, R. and Kwok, S. (1989). *Nature*, **339**, 237.
26. Chou, S. (1990). *J. Infect. Dis.*, **162**, 738.
27. Williams, P., Simmond, P., Yap, P. L., Balfe, P., Bishop, J., Brettle, R., *et al.* (1990). *AIDS*, **4**, 393.
28. Sarkar, G. and Sommer, S. S. (1990). *Nature*, **343**, 27.
29. Cone, R. W., Hobson, A. C., Huang, M. L. W., and Fairfax, M. R. (1990). *Lancet*, **336**, 686.
30. See Walt Disney's *Fantasia*.
31. Eisenstein, B. I. (1990). *New Engl. J. Med.*, **322**, 178.
32. Yamada, M., Hudson, S., Tournay, O., Bittenbender, S., Shane, S. S., Lang, B., Tsujimoto, Y., Caton, A. J., and Rovera, G. (1989). *Proc. Natl. Acad. Sci. USA*, **86**, 5123.
33. Loh, E. Y., Elliot, J. F., Cwirla, S., Lanier, L. L., and Davis, M. M. (1989). *Science*, **243**, 217.
34. Lee, M., Chang, K., Cabanillas, F., *et al.* (1987). *Science*, **237**, 175.
35. Crezenzi, M., Seto, M., Herzig, G. P., Weiss, P. D., Griffith, R. C., and Korsmeyer, S. J. *Proc. Natl. Acad. Sci. USA*, **85**, 4869.
36. Shibata, D., Namiki, T., and Higuchi, R. (1990). *Am. J. Surg. Pathol.*, **14**, 1076.

Note added in proof

A modification to the basic PCR reaction given in *Protocol 4* has been described recently (Erlich, H. A., Gelfand, D., and Sninsky, J. J. (1991). *Science*, **252**, 1643). Termed the 'hot start' technique, the *Taq* polymerase is added after heating the sample to 95°C for 4 minutes, i.e. the initial denaturation. This reduces non-specific primer extension during the initial phase of the reaction.

5

DNA sequencing

MEIRION B. LLEWELYN

1. Introduction

This chapter will describe a protocol for sequencing DNA that allows the reading of between 300 and 500 base-pairs. Part of the difficulty in following protocols for sequencing is that for each of the many stages a number of methods is possible and therefore only a description of the particular methods used in this laboratory will be given here. Alternative protocols, and descriptions of double-stranded and Maxam–Gilbert sequencing are given in another volume of this series (1). In this chapter, the DNA to be sequenced will be assumed to be derived from a polymerase chain reaction (PCR) (see Chapter 4) in view of the increasing interest in sequencing this type of product (2).

The 'chain termination' method of sequencing (3) involves the synthesis of a DNA strand by a DNA polymerase using a single-stranded DNA template. An universal oligonucleotide primer is annealed to the template and this is followed by a period of extension where dTTP, dGTP, dCTP, and radioactively labelled dATP are incorporated into the growing strand. Reactions are terminated by addition of dideoxynucleotides into the strands in four separate reactions, one each for ddTTP, ddGTP, ddCTP, and ddATP (dideoxynucleotides lack a 3' hydroxyl group and cannot form phosphodiester bonds). This results in the production of a family of DNA fragments of different lengths depending on the site of incorporation of the dideoxynucleotide. High-resolution gel electrophoresis separates the strands, and because of the incorporation of radioactive dATP they can be visualized by autoradiography. If the four termination reactions are run next to each other on the gel, then the sequence can be read directly from the order of the ladder of bands (4).

This method depends on the production of single-stranded copies of cloned DNA. A convenient way to achieve this is to use M13 cloning vectors (5). Double-stranded foreign DNA can be cloned into the double-stranded replicative form (RF) 'phage DNA and, upon transformation, only one of the two strands is packaged into the viral coat. The polylinker sites of all M13mp vectors are derived from a similar sequence and it is thus possible to use a single universal primer to initiate extension on all these vectors. Cloning the

PCR products into M13 vectors is essential when a mixture (for example a family) of genes is amplified: each 'phage will contain one insert. Direct sequencing of such a mixture would result in numerous bands in many lanes of a sequencing gel. However, when only one gene is amplified, sequencing can be carried out directly, without initial subcloning.

2. Preparation of the PCR product

For PCR, suitable primer design relies on knowledge of the sequences at the 5'- and 3'-ends of the region of DNA to be amplified. However, for cloning the product into the M13 polylinker site, the primers also need to have selected restriction sites incorporated into their 5'-ends. Thus, PCR amplification followed by restriction digestion of the product allows cloning into M13 cut with the same enzymes. It is useful to incorporate different restriction sites at each end, as the cut M13 will not reanneal to itself and the insert will be in a known orientation (this is known as 'forced cloning'). When designing primers, the following points are vital.

(a) Internal sites. It is important to check that the family of genes to be amplified does not have a conserved internal restriction site that has been incorporated into the primers.

(b) Overhang. After designing the restriction site, the primers must be extended at the 5'-ends, otherwise the enzymes will not cut. The length of the tail needed for each enzyme and the duration of cutting is given in the New England BioLabs Catalogue.

A suitable procedure for use with primers containing *Eco*RI and *Hin*dIII restriction sites is given in *Protocol 1*.

Protocol 1. Purification and restriction digestion of the PCR product

The protocol given takes as an example a restriction digest using *Eco*RI and *Hin*dIII but it can be adapted to any enzymes. This requires the use of primers containing *Eco*RI and *Hin*dIII restriction sites.

Materials

- 10 × medium salt buffer: 0.1 M Tris–HCl, pH 7.5, 0.1 M MgCl$_2$, 10 mM DTT, 1 mg/ml bovine serum albumin (BSA), 50 mM NaCl
- 10 × Tris–borate buffer (TBE): 108 g Tris base, 55 g boric acid, 40 ml 0.5 M EDTA, pH 8.0, made up to 1 litre with deionized water.
- 50 × Tris–acetate buffer (TAE): 242 g Tris base, 57.1 ml glacial acetic acid, 100 ml 0.5 M EDTA pH 8.0, made up to 1 litre with deionized water.

- 3 M sodium acetate, pH 7.0
- Ice-cold ethanol
- Tris–EDTA saturated phenol: saturate with 10 mM Tris–HCl, pH 8.0, 1 mM EDTA (see Chapter 1 for saturation protocol)
- Chloroform: isoamylalcohol (24:1; v/v) mixture
- *Hin*dIII and *Eco*RI restriction endonucleases
- Geneclean® II kit (Bio 101, California, USA)

Method

1. At the end of PCR cycling (see Chapter 4), run 5 μl of the 50 μl PCR product on a 1% agarose gel containing 0.5 μg/ml ethidium bromide and check the length of the product by reference to a set of standards on another track (see Chapter 1).

2. Transfer the remaining 45 μl of product to a 1.5 ml microcentrifuge tube and add 255 μl of water.

3. De-proteinize by adding 300 μl of TE-saturated phenol and vortexing thoroughly.

4. Spin at 1000 *g* in a microcentrifuge for 2 min and transfer the upper aqueous phase, free of any interface material to a clean tube.

5. Add 300 μl of phenol/chloroform mixture (1:1) and repeat the extraction.

6. To the (approximately) 250 μl aqueous phase in a clean tube add 30 μl 3 M Na acetate, pH 7.0 and 900 μl (3 vol.) of pure cold ethanol, chill on dry ice for 20 min and pellet the DNA in a microcentrifuge at 1000 *g* at 4°C for 20 min.

7. Wash the pellet once with 70% ethanol, dry it under vacuum and dissolve it in 34 μl water.

8. To the 34 μl water add:
   4 μl 10 × medium salt buffer
   1 μl *Hin*dIII
   1 μl *Eco*RI

9. After 12 h incubation at 37°C, run the digest on a 1.5% low-melting-point agarose gel (stained with ethidium bromide) in TAE buffer to separate away small fragments.

10. Cut out a gel slice containing the desired fraction over a long-wave UV illuminator and place in a pre-weighed 1.5-ml microcentrifuge tube.

11. The DNA can be recovered easily from the slice by adsorption on to silica particles using a Geneclean® II kit (the procedure takes about 20 min). Re-dissolve the Geneclean® product in 20 μl of water.

3. Preparation of M13 vector DNA for cloning

The vector used in our sequencing method is M13K19 (6), which has been constructed from M13mp19. The vector M13K19 contains an *Eco*K site in the polylinker (as well as four additional unique blunt sites). If a vector with an intact polylinker is transformed into an *Eco*K positive strain of *Escherichia coli* (we use BMH 71–18 (7) available from the American Type Culture Collection, Rockville, Maryland, USA), it is cut by the bacterium. This helps eliminate the problem of background. When an M13mp vector is used, recombinant clones can be differentiated from background vector by employing blue/white selection using media containing 5-bromo-4-chloro-3-indolyl-β-galactoside (X-gal) (1). *Protocol 2* applies to the preparation of any 'phage vector, and is very convenient in that it does not require caesium chloride purification. It is assumed that a small amount of vector RF (acquired commercially) has been transformed into an *Eco*K negative strain of *E. coli* such as TG1 and plaques have formed on a plate (see *Protocol 4*).

Protocol 2. Preparation of replicative form (RF) using lithium chloride (50 ml preparation)

Materials
- 2 × TY medium: 16 g bacto-tryptone, 10 g bacto-yeast extract, 5 g NaCl made up with 900 ml deionized water, pH adjusted to 7.0 using NaOH, and volume made up to 1 litre with deionized water. Sterilize by autoclaving
- GTE solution: 50 mM glucose, 25 mM Tris–HCl, pH 8.0, 10 mM EDTA
- 0.2 M NaOH, 1% sodium dodecyl sulphate (SDS) (store in plastic *not* glass)
- 10% SDS
- Potassium acetate solution: 29.4 g potassium acetate plus 11.5 ml glacial acetic acid in 100 ml water
- Propan-2-ol
- TE buffer: 10 mM Tris–HCl, pH 7.5, 1 mM EDTA
- 4.4 M lithium chloride
- DNase-free RNase (Boehringer Mannheim, Germany)
- Dry ice

Method
1. Toothpick a plaque of 'phage into 10 ml of 2 × TY and shake at 37°C.
2. 1 h later, inoculate 10 ml of 2 × TY in another tube with an overnight growth of TG1, and shake at 37°C for 2 h.

3. Mix the contents of both tubes, make up to 50 ml with 2 × TY and grow at 37°C for a further 4 h to allow the 'phage to infect the TG1 cells.

4. Spin at 1000 g in a 50-ml disposable tube (Falcon, Becton Dickinson, Oxnard, California) for 10 min at 4°C.

5. Re-suspend the cells (on ice) in 4 ml ice-cold GTE.

6. Add 8 ml 0.2 M NaOH, 1% SDS. Shake to mix well and chill on ice for 5 min.

7. Add 6 ml potassium acetate solution, shake well and leave on ice for 10 min.

8. Spin at 1000 g for 15 min at 4°C.

9. Decant the supernatant (containing circular DNA) into another 50-ml tube, discarding the pellet and floating precipitate (chromosomal DNA and cell debris). Add 17 ml of propan-2-ol, and place in dry ice for 15 min.

10. Spin at 1000 g for 15 min, and discard the supernatant. Invert the tube for 5 min and wipe the inside of the tube with a paper towel.

11. Re-suspend the pellet in 2 ml TE buffer, and add 2.5 ml of 4.4 M lithium chloride to precipitate the RNA. Chill on ice for 10 min.

12. Spin at 1000 g for 15 min, add 10 ml of absolute ethanol to the supernatant and leave at room temperature for 15 min.

13. Spin at 1000 g for 15 min. Discard the supernatant and wash the pellet with 70% ethanol.

14. Resuspend the pellet in 400 µl of TE buffer and transfer to a 1.5-ml microcentrifuge tube.

15. Add 2 µl (4 Units) of DNase-free RNase and incubate at 37°C for 15 min.

16. Add 20 µl of 10% SDS and place in a 70°C water bath for 10 min (this destroys a magnesium-dependent DNase that can co-precipitate with DNA).

17. Extract twice with phenol and once with phenol:chloroform (1:1) as described in *Protocol 1* (and Chapter 1).

18. Precipitate the DNA with sodium acetate and ethanol as in *Protocol 1*, but at room temperature or the SDS with also precipitate.

19. Wash the DNA with 70% ethanol, pellet it by centrifugation at 12000 g for 5 min and re-suspend it in 50 µl of TE buffer.

20. Determine the concentration of DNA by UV absorption (see Chapter 1) and adjust to 1 µg/µl with TE buffer.

Having obtained pure double-stranded RF, the next step is to perform a restriction digest using the same enzymes as those chosen for the PCR primers. It is best to cut with one enzyme first, checking on an agarose gel that the RF has linearized, then proceed with the second enzyme. Following digestion (there is no need to use the extended times used for PCR primers) the cut RF should be purified on a low-melting-point TAE gel, as in *Protocol 1*, to remove the excised polylinker. Note that this makes treatment with calf intestinal alkaline phosphatase unnecessary. Determine the concentration of purified M13 by running it on a TBE gel next to a set of standards. Adjust the concentration to 20 ng/μl (redissolving the Geneclean® product in 20 μl of water is usually sufficient). The cut PCR product and M13 vector are now ready for legation.

4. Ligation of vector and PCR product

Most reference works describe this stage using a large number of differential equations. The procedure described in *Protocol 3* is simple, reliable and requires no mathematics.

Protocol 3. Ligation of vector and insert

Materials
- 10 × ligation buffer: 500 mM Tris–HCl, pH 7.6, 100 mM $MgCl_2$, 100 mM DTT, 500 μg/ml BSA
- 10 mM rATP in water (no need to adjust the pH)
- T4 DNA ligase (New England Biolabs, Massachusetts, USA. Note that this is now supplied with buffer containing ATP)

Method
1. To each of three 0.5-ml microcentrifuge tubes add 1 μl each of 10 × ligation buffer, 10 mM rATP, Genecleaned® vector, and DNA ligase.[a]
2. To the first tube add 1 μl of Genecleaned® PCR product, to the second 3 μl of PCR product and to the third 5 μl of PCR product. Make each up to 10 μl with water.
3. Place in a 15 °C water bath for at least 3 h (overnight is equally acceptable).

---

[a] Ligation controls are necessary to pin-point the source of any problems. These are: vector alone and vector and ligase without insert. A vector without an *Eco*K site should be used to assess the transformation efficiency of the competent cells (see *Protocol 4*).

---

5. Isolation of single-stranded recombinant 'phage

The double-stranded recombinant M13 DNA must be converted into a single-stranded template for sequencing. In order to produce a single-stranded 'phage, the double-stranded DNA must first be introduced into *E. coli* (rendered competent by the procedure in *Protocol 4*) by transformation. It is useful to prepare aliquots that can be stored in liquid nitrogen for up to three months without loss of transformation efficiency, and the procedure described in *Protocol 4* is based on that of Hanahan (8).

Protocol 4. Preparation of frozen competent *E. coli* and generation of M13 plaques

Materials
- SOB medium: 20 g bacto-tryptone, 5 g bacto-yeast extract, 0.5 g NaCl in 950 ml deionized water. Add 10 ml of a 250 mM solution of KCl, adjust pH to 7.0 with NaOH and make up to 1 litre. $MgCl_2$ and $MgSO_4$ must be added just before use to a concentration of 10 mM each.
- RF1: 100 mM RbCl, 50 mM $MnCl_2 \cdot 4H_2O$, 30 mM potassium acetate, pH 7.5, 10 mM $CaCl_2 \cdot 2H_2O$, 15% (w/v) glycerol. Adjust the final pH to 5.8 with 0.2 M acetic acid.
- RF2: 10 mM 3-(*N*-morpholeno)propanesulphonic acid (MOPS), pH 6.8, 10 mM RbCl, 75 mM $CaCl_2 \cdot 2H_2O$, 15% (w/v) glycerol. Adjust the final pH to 6.8 with NaOH.
- *E. coli* BMH 71-18

RF1 and RF2 should be sterilized by filtration through a 0.45 μm membrane (Nalgene®, New York) and must be prepared using disposable pipettes and containers; any soap from glassware will destroy bacteria.

Method
1. Pick several colonies of *E. coli* BMH 71-18 from a minimal glucose plate and disperse in 1 ml of SOB to which $MgCl_2$ and $MgSO_4$ have been added as above. Grow at 37°C overnight.
2. Tip these cells into 100 ml of SOB (to which $MgSO_4$ has been added) in a 1-litre soap-free flask and incubate at 37°C with moderate agitation until the absorbance at 595 nm is 0.5 (around 100 min).
3. Pour the culture into two 50-ml polypropylene tubes and chill on ice for 15 min.
4. Centrifuge at 750 g for 15 min at 4°C and drain the pelleted cells thoroughly.
5. Re-suspend the pellet by vortexing moderately in 16 ml (1/3 vol.) of RF1 and leave on ice for 15 min.

Protocol 4. *Continued*

6. Pellet the cells again at 4°C, re-suspend the pellet in 4 ml (1/12.5 vol.) of RF2 and leave on ice for 15 min.

7. Distribute aliquots (0.5–1 ml) into screw-capped tubes, flash-freeze in liquid nitrogen and store in liquid nitrogen or at −70°C.

Cell usage

Materials

- TYE plates: 15 g agar, 8 g NaCl, 10 g bactotryptone, 5 g yeast extract made up to 1 litre with distilled water.
- Molten H top agar: 10 g tryptone, 8 g NaCl, 7 g agar per litre
- Frozen competent cells

Methods

1. Have a water bath ready at 42°C and place in it three glass test-tubes with 3 ml of molten top agar in each.

2. To each tube add 50 μl of an overnight growth of *E. coli* BMH 71-18 in 1 ml of 2 × TY (see *Protocol 2*).

3. Place three TYE plates into a 37°C oven to warm (otherwise the molten agar will solidify too quickly).

4. Thaw the frozen cells in the palm of your hand until just liquid and place them on ice.

5. Into each of three polypropylene (Falcon 2063) round-bottom tubes aliquot 150 μl of competent cells.

6. Add 5 μl of each of the three ligation mixes (see *Protocol 3*) to a tube of competent cells, swirl, and place on ice for 30 min.

7. Heat-shock the cells by placing the tubes in the 42°C water bath for 2 min and replace them on ice.

8. Take a glass tube of molten agar, tip the contents into a tube containing the heat-shocked cells and pour them quickly and evenly over a TYE plate. Repeat for the other two tubes.

9. Allow the agar to solidify for 10 min and place the plates (inverted) in a 37°C incubator. M13-infected (and therefore more slowly growing) cells will be visible as turbid plaques in the background lawn of cells in 8 h. **Clear** plaques could well mean contamination due to T 'phage.

5.1 Preparation of single-stranded DNA

Plaques containing recombinant single-stranded 'phage must now be picked, grown in culture and deproteinized. The TYE plates can be stored in a refrigerator after plaques have become apparent. *Protocol 5* describes the preparation of 20 sequencing templates, enough for two sequencing gels.

Protocol 5. Preparation of single-stranded DNA

Materials

- Polyethylene glycol (PEG): 20% PEG 6000, 2.5 M NaCl.
- 2 × TY medium (see *Protocol 2*)
- Sterile toothpicks

Method

1. Into 100 ml of 2 × TY, pour 1 ml of an overnight growth of *E. coli* BMH 71-18. Aliquot 1.5 ml of this infected medium into each of 20 glass test-tubes.
2. Toothpick a plaque and drop the toothpick into a tube of infected medium. Repeat for 20 plaques, choosing those well-separated from others. Grow at 37°C for 5 h with vigorous shaking; any longer will result in bacterial lysis and contamination with RNA.
3. Transfer the cultures into 1.5-ml microcentrifuge tubes, and spin for 5 min in a microcentrifuge at 12 000 g.
4. Transfer 1 ml of the supernatant (containing free 'phage) into a second 1.5-ml microcentrifuge tube and add 200 μl of PEG. Leave at room temperature for at least 15 min.
5. Centrifuge for 5 min at 12 000 g to precipitate the 'phage, and aspirate the supernatant.
6. Re-spin the pellet briefly and remove all traces of supernatant.
7. Re-suspend the 'phage pellet in 300 μl of water.
8. Extract twice with an equal volume of phenol:chloroform (1:1), precipitate with sodium acetate, wash with 70% ethanol and dry under vacuum (see *Protocol 1*).
9. Re-suspend in 30 μl of water. This is the single-stranded sequencing template.

6. Sequencing reactions

In the original chain-termination method (3), the incorporation of nucleotides into the growing DNA strand was catalysed by the Klenow fragment of *E. coli* DNA polymerase I. However, bacteriophage T7 DNA polymerase (9) confers several advantages over Klenow for sequencing: high processivity and rapid polymerization generates long fragments with very few false terminations. Kits for sequencing have made the procedure very straightforward and are recommended: among those using T7 polymerase are Sequenase® II

(United States Biochemical, Cleveland, Ohio, USA), and kits available from Promega (Madison, Wisconsin) and Amersham (Amersham, UK). The Sequenase® II kit also contains reagents for sequencing with dITP and with manganese buffer, and the basic Sequenase® kit reagents will be described here. The procedure described in *Protocol 6* is performed in microtitre plates and therefore requires a centrifuge with adaptors for spinning microtitre plates (for example, an IEC Centra 4B) and, as termination is performed at 42°C, a heating block capable of holding microtitre plates (for example, a Techne Dri-plate® cycler MW 1, UK). Alternative procedures can be found in ref. 1 and in the manufacturer's instructions provided with the sequencing kits.

Protocol 6. Sequencing reactions

Materials

- 5 × reaction buffer: 200 mM Tris–HCl, pH 7.5, 100 mM $MgCl_2$, 250 mM NaCl
- Enzyme dilution buffer: 10 mM Tris–HCl, pH 7.5, 5 mM DTT, 0.5 mg/ml BSA
- 5 × labelling mixture: 7.5 μM dGTP, 7.5 μM dCTP, 7.5 μM dTTP
- ddG termination mixture: 80 μM dGTP, 80 μM dATP, 80 μM dCTP, 80 μM dTTP, 8 μM ddGTP, 50 mM NaCl
- ddA termination mixture: 80 μM dGTP, 80 μM dATP, 80 μM dCTP, 80 μM dTTP, 8 μM ddATP, 50 mM NaCl
- ddC termination mixture: 80 μM dGTP, 80 μM dATP, 80 μM dCTP, 80 μM dTTP, 8 μM ddCTP, 50 mM NaCl
- ddT termination mixture: 80 μM dGTP, 80 μM dATP, 80 μM dCTP, 80 μM dTTP, 8 μM ddTTP, 50 mM NaCl
- Stop solution: 90% formamide, 20 mM EDTA, 0.05% bromophenol blue, 0.05% xylene cyanol
- Universal-40 sequencing primer (0.5 pmol per μl)
- [^{35}S] dATP αS (1 mCi/37 MBq per 100 μl) (Amersham, UK: not supplied in kits)
- 0.1 M DTT

Method

All additions are made with a 2-μl Hamilton dispenser (PB600) fitted with a 1710 gas-tight syringe and adaptor. Dilute the labelling mixture fivefold at the outset.

1. For each sequencing template, mix in a 1.5-ml microcentrifuge tube 6 μl water, 1 μl universal primer, and 2 μl reaction buffer to give the primer mixture.

2. Label a Falcon 3911 U-bottomed microtitre plate with the number of the clone across the top and TCGA down the left side.

3. Aliquot $2\,\mu l$ of primer mixture into the bottom of each well, add $2\,\mu l$ of sequencing template to the side wall of the appropriate well, spin, cover with Saran® wrap and a microtitre lid, and float the plate in a water bath at 70°C for 5 min. Leave to cool on a bench for 30 min (the primer is now annealed to M13).

4. During the period of cooling, prepare the labelling mixture by adding (for each sequencing template) in a 1.5-ml microcentrifuge tube, $0.5\,\mu l$ $[^{35}S]$ dATP, $1\,\mu l$ 0.1 M DTT, $2\,\mu l$ diluted labelling mixture and $3.5\,\mu l$ water.

5. Label a Techne 96® polycarbonate microtitre plate in the same way as the first microtitre plate, and into wells on the row 'T' aliquot $2\,\mu l$ of ddT termination mix, doing the same for the other three rows. Place the plate on a heating block at 42°C.

6. When the microtitre plate from step 3 has cooled for the 30 min, add to the labelling mixture (per template) $1.77\,\mu l$ of enzyme dilution buffer followed by $0.22\,\mu l$ of Sequenase® II enzyme. (This ensures that the Sequenase® spends the minimum time out of the freezer.)

7. Aliquot $2\,\mu l$ of this labelling mixture on to the side walls of the wells containing primer mix and spin the plate to mix. Start a timer.

8. After 2 min, start pipetting the solutions in the first microtitre plate into their equivalent positions on the pre-warmed polycarbonate microtitre plate on the heating block. Use an ordinary micropipette, changing the tip for each well and do it quickly (remember that the disposable tips are now radioactive).

9. When the last well has been reached, start a timer, and draw stop solution into a pipette tip at the end of the Hamilton syringe.

10. At time 5 min apply $2\,\mu l$ of stop solution to the side wall of each well, and then spin the plate. This can now be stored (with a Falcon microtitre plate lid) in the freezer until required (^{35}S can be stored for one week at -20°C).

Note that the protocol in the Sequenase® II kit explains how to perform the reactions in **tubes** rather than in microtitre plates and will prove useful if the equipment mentioned above is unavailable.

7. Gel electrophoresis

7.1 Preparation of the sequencing gel

The mixtures of DNA strands of varying lengths in each microtitre plate well can be fractionated through polyacrylamide by electrophoresis. In the presence of the initiators ammonium persulfate (AMPS) and *N,N,N',N'*-tetramethylethylenediamine (TEMED), acrylamide produces long chains of polyacrylamide. If *N,N'*-methylene bisacrylamide is included in the monomer mixture, branches are formed that produce bridges between polyacrylamide chains. The polyacrylamide gel described in *Protocol 7* is a **buffer gradient** gel (10) that, by virtue of having a higher buffer salt concentration within the lower one-third of the gel, retards the smaller and thus faster moving fragments. This clearly increases the resolving power of the gel, but if this property is not required it is possible to pour a gel made using only the 0.5 × gel mix described in *Protocol 7*.

There are numerous commercially available gel apparatuses (for example, the International Biotechnologies Inc. Base Runner®, New Haven, Connecticut, USA). 60-cm gel plates and 0.4-mm spacers are convenient, and a 48-well 0.4-mm sharkstooth comb (Bethesda Research Laboratories Gaithersburg, Maryland, USA) allows twelve sequencing reactions to be run per gel. The steps outlined in *Protocol 7* apply to any gel apparatus.

Protocol 7. Preparation of a buffer gradient sequencing gel

Note: gloves must be worn at all times.

Materials

- 40% acrylamide solution: acrylamide powder, 380 g; *N,N'*-methylene bisacrylamide 20 g. Make up to 1 litre with deionized water, add Amberlite MB1 resin and stir for 10 min. Filter through Whatman paper into a dark glass container, and store at 4°C. Warning: highly toxic
- 25% ammonium persulfate (AMPS) (must be prepared fresh each time)
- TEMED (BDH, Poole, Dorset)
- 10 × TBE buffer (see *Protocol 1*)
- Dimethylchlorosilane solution (BDH, UK)

Method

1. Prepare a 5 × TBE, 6% gel mix by adding together 92 g urea, 30 ml 40% acrylamide, 100 ml 10 × TBE and 10 mg of bromophenol blue. Make up to 200 ml with deionized water. Filter through Whatman paper and store at 4°C. This solution lasts for 1 month.

2. Prepare a 0.5 × TBE, 6% gel mix by adding together 23 g urea, 7.5 ml 40% acrylamide, 2.5 ml of 10 × TBE and 25 ml deionized water. Unlike the 5 × mix, this must be made fresh each time. Filter through Whatman paper into a 100-ml beaker.

3. Wash the glass plates and spacers in warm detergent solution and rinse in deionized water. Wipe the plates with ethanol, using Kimwipes®, and then polish them to remove any grease. Repeat with acetone.

4. Smear 2 ml of dimethylchlorosilane solution over the notched plate to silanize it. Clean this surface with deionized water, ethanol and acetone.

5. Lay the larger glass plate on the bench, and arrange the two flexible plastic spacers along both sides. Lay the smaller plate in position over it and form a liquid-tight seal around the edges using yellow electrical tape (3M, BRL). Place two large bulldog binder clips on each side of the mould, one at the bottom and one in the middle.

6. Place 7.5 ml of the blue 5 × TBE mix in a 10-ml beaker.

7. To the 50 ml of 0.5 × TBE mix in the 100 ml beaker, add 65 µl of AMPS and 65 µl of TEMED and swirl.

8. To the 7.5 ml of 5 × TBE mix add 16 µl of AMPS and 16 µl of TEMED and swirl.

9. Draw 30 ml of the 0.5 × TBE mix into a 50-ml syringe and lay this down on the bench.

10. Draw 6.5 ml of the remaining 0.5 × TBE mix into a pipette, and then draw all the 5 × TBE mix into the same pipette. Pull through two bubbles to help mix the solutions.

11. Hold the gel mould at an angle at 45 degrees to the bench, tipped slightly away. Start pouring the contents of the pipette slowly down the far wall of the mould, lowering the mould as the pipette empties.

12. When the pipette is empty and the mould flat, pick up the 50-ml syringe and commence filling the mould as before, liftng it up at the same time. When the mould is full, balance the top end on the 10-ml beaker (this keeps it at a good elevation). Any bubbles in the upper part of the gel can be scooped out using a spacer thinner than 0.4 mm.

13. Place the flat side of the sharkstooth comb in the slot to a depth of around 6 mm, place one pair of bulldog clips in a deep bite along each side of the upper part of the plate, and a second pair to bite into the notched plate from above. Leave to set for at least 1 h. If leaving overnight, cover the top part with a tissue soaked in 1 × TBE and cover the tissue with Saran® wrap.

7.2 Loading and running the gel

A procedure for loading and running the sequencing gel prepared in *Protocol 7* is given in *Protocol 8*.

Protocol 8. Loading and running the gel

1. Wash away any excess polymerized gel from the outside of the mould with deionized water, and remove the bulldog clips. Slowly pull out the sharks-tooth comb whilst deionized water is running over it.

2. Slit the sealing tape at the bottom of the mould with a razor blade, and place the mould in the electrophoresis apparatus.

3. Fill the upper and lower electrophoresis tanks with 1 × TBE and wash away any urea and unpolymerized acrylamide from the top of the gel with a syringe and needle.

4. Insert the sharkstooth comb with the pointed teeth just penetrating the gel and check that they do not leak by applying a few microlitres of blue formamide stop solution from the sequencing kit between the teeth at the extreme ends and middle of the comb. (This is best done using a disposable Multi-Flex® pipette tip mounted on a micropipette.)

5. Put the polycarbonate microtitre plate containing the sequencing reactions on a heating block (or in an oven) at 80°C for 15 min to denature and concentrate the sample (note that a Falcon plate would buckle). Spend this time flushing out each well of the gel with your pipette.

6. Load 2 µl of sample into each well in a predetermined order (we use 'TCGA'). Load all the samples by 30 min or some renaturation will occur.

7. When the gel is loaded, connect to a power supply (we use an LKB 2197 power supply) and run at 45 W for 4 h (for a 60 cm gel). The actual time required for a particular sequence can be adjusted on the basis of the result at 4 h.

7.3 Autoradiography of the gel

The development of the sequencing gel produced by *Protocol 8* requires autoradiography. This is described in *Protocol 9*.

Protocol 9. Autoradiography of the gel

1. When electrophoresis is completed, disconnect the power supply and dispose of the TBE solutions. Remember that the lower chamber fluid is now radioactive.

2. Remove the mould, strip off the electrical tape and lay the mould on the bench with the notched plate uppermost. Insert a razor blade or spatula between the two plates and separate them gently.

3. Transfer the gel and the supporting plate into a shallow bath containing 10% methanol and 10% acetic acid in water to remove urea and fix the gel. Do not agitate.

4. After 15 min take the plate out and tip it gently to remove excess fluid and place on the bench to dry a little for 10 min.

5. Lay a piece of Whatman 3MM chromatography paper, 3 cm longer on all sides than the gel, on top of the gel and press down firmly. Peel back the paper, and the gel will adhere to it. Place, gel uppermost, on another piece of Whatman paper.

6. Put a layer of Saran® wrap on top of the gel and trim the Saran® wrap and Whatman paper to the same size as the gel.

7. Dry the gel under vacuum for 40 min on a gel drier set at 80°C, with the Saran® wrap uppermost.

8. After 40 min, peel away the Saran® wrap from a corner and feel the gel. If the gel is very sticky, it is not dry enough and should be dried for a further 10 min.

9. Peel off the Saran® wrap and run a Geiger counter over the gel. With experience it is possible to judge from the counts the length of time needed for autoradiography.

10. Put the gel in contact with a piece of film (Kodak XAR or Fuji RX) in a film cassette and expose at room temperature without an image intensifier for 16–48 h.

11. Develop the autoradiograph and read the sequence of DNA.

---

8. Direct sequencing of PCR products

When looking for mutations in specific genes that are present in the genome in only one or two (allelic) copies (for example, mutations responsible for sickle-cell trait) then it is much more convenient to sequence the PCR product directly. If two alleles are amplified and sequenced, the sites of mutation will appear as ambiguous bands on the gel, and if the sequence of the normal gene is known the mutations can be identified. Where there are no alleles, for example in mitochondrial DNA, then the actual sequence may be read directly. (Note that sequencing kits are not licensed for human diagnosis.) There are several ways to accomplish direct sequencing (11): for example a radiolabelled primer can be added to the double-stranded PCR product after spin dialysis or gel purification to remove PCR primers and unincorporated nucleotides. In asymmetric PCR sequencing, an excess of single-stranded PCR product is generated by using different molar ratios of the two PCR primers. The concentration of one of the primers becomes limiting and the other primer generates single-stranded product. Again, excess PCR primers and nucleotides need to be removed before sequencing.

We use a method whereby one of the PCR primers is biotinylated, and pure single-stranded DNA is produced by adding streptavidin coated magnetic beads to the PCR product and placing the mixture in a magnet. It is then simple to wash away the second DNA strand, the PCR primers and excess nucleotides (2). The biotinylation of the primer is best accomplished by incorporating a biotinylated nucleotide at the 5'-end during synthesis. The procedure is described in *Protocol 10*.

Protocol 10. Direct PCR sequencing using magnetic beads

Materials
- Streptavidin-coated magnetic beads (Dynabeads® type M-280, Dynal (UK) Ltd, Wirral, UK)
- Magnet for holding 1.5-ml microcentrifuge tubes (Dynal MPC®-E)
- PBS buffer: 10 mM phosphate, 150 mM NaCl, pH 7.2
- PBS/BSA buffer: 4.5 ml PBS and 0.5 ml 10 mg/ml BSA
- 0.15 M NaOH

Method
1. Perform the PCR reaction in the normal way (see Chapter 4) in a 50 μl volume using 5 pmol of each primer, as excess biotinylated primer will reduce the efficiency of the beads.
2. Aliquot 20 μl of Dynabeads (10 mg/ml) into a clean microcentrifuge tube and place the tube in the magnet for 30 sec. This will cause the beads to aggregate on to the side wall.
3. Add 1 ml of PBS/BSA buffer and remove from the magnet to mix gently.
4. Replace in the magnet, remove the liquid and repeat five times to wash out azide that is included as an antibacterial agent. Finally, re-suspend in 20 μl of PBS/BSA buffer.
6. Add the 20 μl of re-suspended beads to the PCR product in a clean 1.5 ml microcentrifuge tube (there is no need to remove the paraffin) and leave for 15 min.
7. Place the tube in the magnet for 30 sec, remove the liquid and add 50 μl 0.15 M NaOH. Remove the tube from the magnet and let stand for 5 min.
8. Repeat step 7.
9. Wash once with 0.5 ml water and finally re-suspend in 10 μl water. This may be treated as a sequencing template (see *Protocol 6*) and is enough for one set (four lanes) of reactions.

It is worthwhile starting with 2 µl of template in each microtitre plate well as outlined in *Protocol 6*, but it may prove necessary to adjust the volumes of template and primer. The first strategy, should the sequence be faint throughout, is to increase the amount of sequencing primer, which can be the unlabelled PCR primer. Alternatively, an internal primer may give fewer false terminations. Start at 0.5 pmol/µl of primer and increase by a factor of 10 in other sequencing reactions.

Acknowledgements

The author is grateful to Dr Greg Winter and Michael Tristem for helpful suggestions and advice.

References

1. Howe, C. J. and Ward, E. S. (ed.) (1989). *Nucleic acids sequencing: a practical approach*. IRL Press, Oxford.
2. Gyllenstein, U. B. (1989). *Biotechniques*, **7**, 700.
3. Sanger, F., Nicklen, S., and Coulson, A. R. (1977). *Proc. Natl. Acad. Sci. USA*, **74**, 5463.
4. Watson, J. D. (1987). In *Molecular biology of the gene* (ed. J. D. Watson, N. H. Hopkins, J. W. Roberts, J. A. Steitz, and A. M. Weiner), p. 277. Benjamin/Cummings Co., Menlo Park, California.
5. Messing, J. (1983). In *Methods in enzymology*, Vol. 101 (ed. R. Wu, L. Grossman, and K. Moldave), p. 20. Academic Press, London.
6. Carter, P., Bedouelle, H., Waye, M. Y., and Winter, G. P. (1985). *Oligonucleotide site-directed mutagenesis in M13*. Anglian Biotechnology Ltd, Hawkins Road, Colchester, UK.
7. Gronenborn, B. (1976). *Mol. Gen. Genet.*, **148**, 243.
8. Hanahan, D. (1983). *J. Mol. Biol.*, **166**, 557.
9. Tabor, S. and Richardson, C. C. (1989). *J. Biol. Chem.*, **264**, 6447.
10. Biggin, M. D., Gibson, T. J., and Hong, G. F. (1983). *Proc. Natl. Acad. Sci. USA*, **80**, 3963.
11. Sambrook, J., Fritsch, E. F., and Maniatis, T. (ed.) (1989). *Molecular cloning, a laboratory manual*. Cold Spring Harbor Press, Cold Spring Harbor, NY.

6

Determination of genital human papillomavirus infection by consensus PCR amplification

HEIDI M. BAUER, CATHERINE E. GREER,
and M. MICHELE MANOS

1. Introduction

Genital human papillomaviruses (HPVs) are sexually transmitted, and are believed to be a major contributing factor in the aetiology of several anogenital cancers, including cervical squamous cell cancers and adenocarcinomas, vulvar carcinomas, and penile and anal cancers (for review see refs. 1, 2). Thus the sensitive and accurate detection and typing of genital HPVs in normal, dysplastic, and cancer tissue samples is essential to understanding the role of these viruses in the various pathologies they effect. Furthermore, as that understanding improves, the screening of patients for particular HPVs may provide valuable clinical information.

Also known as wart viruses, the HPV group contains over 65 distinct cutaneous and mucosal virus types (3). Over 20 of these HPV types infect genital and anal mucosa. HPV types 6 and 11 are commonly associated with benign anogenital warts, while types 16 and 18 are found in a majority of carcinomas. Other types (for example, HPVs 31, 33, 39, and 45) are also reported to be associated with CIN or carcinoma. Since an *in vitro* propagation system is not available for diagnosing HPV infection, methods for the direct detection of the viral DNA are necessary.

1.1 Overview of the PCR system

Polymerase chain reaction (PCR) DNA amplification is the most sensitive method for nucleic acid detection currently available. The HPV detection systems described here are highly sensitive DNA amplification methods that use degenerate or mixed consensus primers to amplify a broad spectrum of HPV types (4–6). One set of HPV primers targets the highly conserved late region 1 (L1) which encodes a viral capsid protein. In order to detect the myriad of HPV types amplified, this system uses a set of long (400 base-pair) generic probes to detect HPV amplification products (7). This system can be used to amplify and detect HPV types 6, 11, 16, 18, 26, 31, 33, 35, 39, 40, 42, 45, 51, 52, 53, 54, 55, 56, 57, 59, and at least 25 yet unidentified HPV types.

Another set of PCR primers targets the early region E6 which is most likely to be retained if viral DNA integration has occurred in the infected cells. This system uses a mixture of consensus primers and is known to amplify HPV types 6, 11, 16, 18, 31, 33, 39, 42, 45, and 52 (6). We have yet to determine whether these primers will amplify as broad a spectrum as the L1 consensus primers. Since a generic HPV E6 probe is as yet unavailable, HPVs are identified by using type-specific oligonucleotide probes.

Clinical specimens undergo crude preparation to free the DNA, and are subsequently amplified using the consensus HPV primers as well as human β-globin primers. The co-amplification of human β-globin serves as an internal control for specimen integrity (see also Chapter 4.)

PCR products may be analysed first using polyacrylamide gel electrophoresis and ethidium bromide staining. Definitive detection and typing of the viral DNA is performed by dot blot hybridization. The presence of an HPV L1 amplification product is determined by hybridization with a generic probe mixture. HPV types are distinguished using type-specific oligonucleotide probes. This system affords the detection of as few as ten copies of HPV DNA in a background of 1000 human cell DNA equivalents per reaction.

1.2 Applications

These HPV detection systems can be applied to many different kinds of clinical material including genital swab samples, scrapes, lavages, biopsy specimens, and paraffin-embedded tissues (PETs). Samples can be collected non-invasively from oral, anal, or genital mucosal tissue as well as epidermal tissue. Plasmid and cell line DNA can also be analysed using this method.

The application of this method to fresh clinical material allows medical scientists and epidemiologists to study the relationship of HPV infection to disease. Molecular epidemiological studies can be undertaken to determine the prevalence of the virus as well as the risk factors that predict infection.

The application of PCR technology to studies of fixed, paraffin-embedded tissues provides molecular biologists and pathologists with a powerful tool to analyse vast collections of archival tissues (see also Chapter 4). This detection method is useful for retrospective studies concerning the correlation of histopathological changes in tissues with the presence of HPV. In addition, the role of HPV in the progression to malignancy can be studied and the clinical utility of HPV typing can be assessed.

2. Preparation of clinical samples

Because of the extreme sensitivity of the DNA amplification methods, it is critical to minimize the amount of sample manipulation. The following methods provide a crude cell lysis that is suitable for PCR. Contamination of clinical samples can occur in three ways during handling: sample-to-sample contamination, introduction of PCR product DNA into the sample, and introduction of HPV-containing plasmid or cell line DNA. To avoid the latter two possibilities, sample preparation should be carried out in an area that is completely free of HPV DNA and PCR product DNA. To avoid the former possibility, great care in handling (opening and closing tubes, for example) must be taken. In addition, only sterile disposable supplies should be used and gloves should be changed frequently. Equipment and working surfaces should be treated frequently with bleach or hypochlorite. These 'PCR clean' techniques should also be used at the clinic or hospital where samples are collected.

Controls are essential for monitoring contamination, reproducibility, and sensitivity of the assay. Negative controls that mimic specimens (that is, containing human cells or human DNA without HPV DNA) should be interspersed throughout sample manipulation. Positive controls with low amounts of HPV DNA should be processed to determine the analytical sensitivity.

2.1 Genital swab or scrape

Since the sample preparation methods are crude, the collection buffer is an important consideration. The ideal buffer would inactivate infectious agents in the sample, yet allow for PCR amplification. Unfortunately, the *Taq* DNA polymerase enzyme is inhibited by various chemicals used to inactivate samples. One commercial collection buffer that has been used in our laboratory is packaged as part of the ViraPap™ HPV detection kit (Digene Diagnostics, Inc., USA). The kit contains materials for the dot blot analysis of samples using radiolabelled RNA probes, but it is possible to purchase only the collection tubes.

Protocol 1. Sample preparation from swabs collected in ViraPap™ collection tubes

Materials
- ViraPap™ collection tubes (Digene Diagnostics, Inc., USA)
- 5 M ammonium acetate
- 100% Ethanol
- ViraPap™ kit blue digestion solution
- TE buffer: 10 mM Tris–HCl [pH 7.5], 1 mM EDTA

Protocol 1. *Continued*

Method

1. To each sample, add 2 drops of the blue digestion solution supplied in the ViraPap™ kit. Incubate at 37°C for 1 h. After digestion, samples may be stored at −20°C, or carried through the following steps.
2. Label 1.5-ml microcentrifuge tubes and add 56 μl of 5 M ammonium acetate and 340 μl of 100% ethanol to each tube.[a]
3. Remove 100 μl of the digested sample with a disposable pipette and add slowly to the tube containing ammonium acetate and ethanol. Close the cap and mix by inversion.
4. Chill to precipitate the DNA at −20°C overnight or at −70°C for 15 min.
5. Pellet the DNA in a microcentrifuge at 12 000 g for 15 min. Carefully draw off and discard the supernatant without disturbing the pellet.
6. Dry the pellet at 55°C for 15 min.
7. Suspend the DNA in 50 μl TE and store at −20°C.
8. Use 10 μl of prepared sample per 100 μl PCR reaction. **Always** spin the sample tubes briefly before opening.

---

[a] The ammonium acetate and ethanol can be pre-mixed. For 80 ml, mix 66.8 ml 100% ethanol and 13.2 ml 5 M ammonium acetate. Use 400 μl per 100 μl of sample.

As described in *Protocol 1*, samples collected in commercial buffers may require alcohol precipitation. More extensive purification methods (for example, phenol extraction or filtration) are possible but not favourable since they involve extensive manipulations that increase the risk of contamination. Saline and other physiological buffers are ideal for PCR, but the clinical material may constitute a biological hazard.

2.2 Saline cervicovaginal lavage

Specimens collected in the form of a saline lavage require only protein digestion prior to amplification. These samples are vulnerable to enzymatic degradation and must be kept frozen. In addition, any infectious agents present remain viable. Thus, they must be handled as a biohazard. Cervicovaginal lavages typically contain 1–10 μg of DNA. To avoid cross-contamination, it is important to minimize the handling of these samples at the clinic at which they are collected.

The following protocol can be applied to lavages, swabs, or scrapes collected in saline. If specimens are collected in PBS, beware that amplification reactions containing more than 10% PBS are inhibited.

Protocol 2. Sample preparation from saline lavage

Materials

- Digestion buffer: 50 mM Tris–HCl [pH 8.5], 1 mM EDTA
- Proteinase K, 20 mg/ml in sterile distilled water
- 10% (v/v) Laureth-12 (Mazer Chemical, Gurney, Illinois, USA) or Tween-20 non-ionic detergent

Method

1. Prepare 2 × digestion solution containing 400 µg/ml proteinase K and 2% (v/v) Laureth-12 (or Tween-20) in digestion buffer (for example, for 0.5 ml, mix 390 µl digestion buffer, 100 µl 10% Laureth-12, and 10 µl 20 mg/ml proteinase K).
2. Aliquot 50 µl of 2 × digestion solution to labelled 1.5-ml microcentrifuge tubes.
3. Remove 50 µl of sample and add to a tube containing 2 × digestion solution. Incubate at 55°C for 1 h. Spin in a microcentrifuge at 12 000 g briefly.
4. Inactivate the protease by incubating the tubes at 95°C for 10 min. Store at −20°C.
5. Use 10 µl of digested sample per 100 µl PCR reaction. **Always** spin the sample tubes briefly before opening.

2.3 Paraffin-embedded tissues (PETs)

The ability of PETs to serve as amplification templates is determined in part by the fixative that was used, the length of fixation time, and the age of the specimen (8). In some cases the DNA is damaged and only very small fragments can be amplified. Thus, it is especially important with PETs to include internal PCR control amplifications that not only demonstrate the presence of human tissue, but the amplifiability of large fragments of DNA. The amplification of PETs can be challenging. However, many of the difficulties can be overcome. We highly recommend further reading in this area (refs. 8, 9, and see Chapter 4).

Protocol 3. Sample preparation from paraffin-embedded tissues

Materials

- Octane or xylene
- 100% Ethanol

Protocol 3. *Continued*

- Acetone (HPLC-grade)
- Digestion buffer: 50 mM Tris–HCl [pH 8.5], 1 mM EDTA
- 10% Laureth-12 or Tween-20 non-ionic detergent
- Proteinase K, 20 mg/ml in sterile distilled water

Method

1. Prior to sectioning each block, thoroughly clean the microtome surface and blade with xylene or octane then ethanol to avoid cross-contamination of samples. Change gloves between every block.

2. Cut 10–20 μm sections from each block and, using sterile toothpicks, place each section in a sterile 1.5-ml microcentrifuge tube. A negative control block (e.g. appendix tissue) should be interspersed every 10 blocks to monitor possible contamination.

3. Extract the section with 1 ml of octane. Mix on a rotating or rocking shaker at room temperature for 15–30 min or until the paraffin is solubilized.

4. Pellet the tissue by centrifugation at 12 000 g for 2 min. Carefully remove the solvent with a disposable pipette, without removing any tissue. If paraffin is still present, repeat steps 3 and 4.

5. Suspend in 0.5 ml 100% ethanol. Centrifuge for 2 min at 12 000 g and remove as much ethanol as possible.

6. Add 10 μl of acetone and dry (tubes open) in a 55 °C heating block or oven.

7. Prepare digestion solution having a final concentration of 200 μg/ml proteinase K and 1% Laureth-12 (or Tween-20) in digestion buffer.

8. Suspend the dry pellet in 50–250 μl digestion solution depending on the size of the section. Incubate for 2–3 h at 55 °C, or overnight at 37 °C. Spin briefly at 12 000 g.

9. Heat-inactivate the protease by incubating the tubes at 95 °C for exactly 10 min.

10. Prior to PCR, spin any debris to the bottom of the tube. Amplify at least two different sample volumes (for instance, 1 and 10 μl) per 100 μl PCR reaction (see *Protocol 4*).

---

3. Nucleic acid amplification

The polymerase chain reaction uses oligonucleotide primers to direct the synthesis of a specific fragment of DNA (10, 11). Using the thermostable enzyme *Taq* DNA polymerase, the reaction can be temperature cycled many times through strand denaturation, primer annealing, and polymerization. This cycling results in the exponential accumulation of the specific fragment of DNA (see Chapter 4). Because of its ability to amplify less than ten copies of target DNA per reaction, this method is highly sensitive and prone to contamination.

3.1 Avoiding PCR contamination

Because of the extreme sensitivity of PCR, many precautions must be taken to avoid contamination (12). For example, in order to reduce contamination from aerosols remaining in pipetmen, it is necessary to use positive-displacement pipette tips or plugged pipette tips when pipetting samples and PCR reagents.

It is strongly suggested that an area such as a bacterial hood with a UV light be dedicated to the preparation of PCR reagents and amplification of samples. If a fume or laminar flow hood is used, the fan should be turned off, since the air flow is a potential source of sample contamination.

All pipetters, pipettes, bulbs, tube racks, etc., must be designated for PCR preparation only and kept in this hood or clean area at all times. These instruments should never be used to pipette PCR product DNA, plasmid DNA or other potentially contaminating DNA sources. Surfaces and instruments should be cleaned frequently with bleach or hypochlorite. Minuscule amounts of amplified target sequences can cause carry-over contamination and invalidate an entire study.

3.2 HPV L1 and β-globin amplification

The procedures outlined here are for the simultaneous amplification of an HPV L1 fragment (450 bp) and human β-globin fragment (268 bp) (*Protocol 4*). An additional larger β-globin product (536 bp) can be amplified in parallel when analysing PET samples to determine if the PET DNA is sufficiently intact to produce a product larger than the 450 bp L1 product (see *Protocol 5*).

Positive and negative control DNAs are essential for interpreting sample data. Cell lines derived from cervical cancers that contain HPV are useful as positive controls: SiHa cells contain one copy of HPV 16 per cell, HeLa cells contain 10–50 copies of HPV 18 per cell, and CaSki cells contain 500 copies of HPV 16 per cell. Human cell lines (such as K562) and fresh tissue that do not contain HPV are useful negative controls.

Protocol 4. Co-amplification of HPV L1 and human β-globin genes

Materials
- Sterile, glass-distilled water
- 10 × *Taq* PCR buffer: 100 mM Tris–HCl [pH 8.5], 500 mM KCl, 40 mM $MgCl_2$
- 10 mM dNTP mixture: 10 mM each of dATP, dCTP, dGTP, dTTP

Protocol 4. *Continued*

- *Taq* DNA polymerase, 5 U/μl
- Light mineral oil
- PCR primers
    HPV primers: MY09, MY11, each at 50 μM (see *Table 1*)
    β-Globin (268 bp) primers: GH20, PC04, each at 50 μM (see *Table 1*)

Method

1. Prepare a PCR master mixture for the number of reactions required by adding in order: water, buffer, dNTPs, *Taq* DNA polymerase, and primers.
2. Each PCR contains the following:
    - Sample DNA   1–10 μl
    - 10 × PCR buffer   10 μl
    - 10 mM dNTP mix   2 μl
    - *Taq* DNA polymerase   0.5 μl
    - MY09 @ 50 μM   1 μl
    - MY11 @ 50 μM   1 μl
    - GH20 @ 50 μM   0.1 μl
    - PC04 @ 50 μM   0.1 μl
    - Water   to 100 μl

    In terms of concentration, each reaction contains 10 mM Tris–HCl, 50 mM KCl, 4 mM $MgCl_2$, 200 μM of each dNTP, 2.5 U *Taq* DNA polymerase, 0.5 μM of each HPV primer, and 50 nM of each β-globin primer.
3. Aliquot 100 μl of mineral oil and 90–99 μl master mix to each labelled 0.5-ml microcentrifuge tube for PCR. The final reaction volume can vary up to 10% without affecting amplification efficiency.
4. Add the sample DNA last. Cap each tube after the addition of DNA.
5. Centrifuge the samples in a PCR-clean microcentrifuge for 10 sec to bring aqueous liquid under the oil.
6. Program a thermal cycler to cycle as follows:[a]
    (a) 35 cycles of: 95°C for 1 min denaturation
                                 55°C for 1 min annealing
                                 72°C for 1 min extension
    (b) Extend: 72°C for 5 min
    (c) Soak: 15°C

[a] When amplifying DNA prepared from PETs, 40 cycles of amplification are required each with a 2-min extension at 72°C as in *Protocol 5*.

Table 1. PCR primer sequences for the HPV L1 and β-genes[a]

Name	5′→3′ nucleotide sequence	Target	Strand	Product size
MY11[b]	GCMCAGGGWCATAAYAATGG	HPV L1	+	450
MY09[b]	CGTCCMARRGGAWACTGATC	HPV L1	−	
GH20[b]	GAAGAGCCAAGGACAGGTAC	β-globin	+	268
PC04[b]	CAACTTCATCCACGTTCACC	β-globin	−	
KM29	GGTTGGCCAATCTACTCCCAGG	β-globin	+	536
RS42	GCTCACTCAGTGTGGCAAAG	β-globin	−	

[a] Degenerate code: M = A + C, R = A + G, W = A + T, Y = C + T
[b] Purchasable through Perkin Elmer-Cetus, Norwalk, Connecticut, USA.

Protocol 5. Amplification of the large β-globin fragment

Materials

Same as *Protocol 4*, except PCR primers

β-Globin (536 bp) primers: KM29, RS42, each at 100 μM (see *Table 1*).

Method

1. Use the PCR preparation guide-lines and reaction conditions outlined in *Protocol 4* with the following primer substitutions:
   - KM29 @ 100 μM   0.1 μl
   - RS42 @ 100 μM   0.1 μl
2. Temperature cycle as follows:[a]

(a) 40 cycles of:	95°C for 1 min denaturation
	55°C for 1 min annealing
	72°C for 2 min extension
(b) Extend:	72°C for 5 min
(c) Soak:	15°C

[a] When amplifying DNA prepared from fresh clinical material, 35 cycles of amplification are required each with a 1-min extension at 72°C as in *Protocol 4*.

3.3 HPV E6 amplification

The E6 amplification system was designed to amplify a smaller fragment (240 bp) from a different region of the HPV genome (6). When used in conjunction with the L1 system, the E6 analysis can provide confirmational typing data for each sample. In addition, the amplification of the smaller E6 product may be favoured in samples containing highly degraded DNA. In the rare event that L1 sequences are altered or disrupted, the E6 amplification can be used to detect HPV.

The E6 amplification system is highly specific, but has a somewhat narrower type spectrum. Instead of relying on one pair of degenerate primers, it uses a mixture of positive and negative strand primers that target the same region of the different HPV types (see *Table 2*).

If the E6 system is used alone, it is necessary to include an internal control amplification of human β-globin. The amplification can be performed either in the same reaction or in parallel. The E6 product and the small (268 bp) β-globin product are difficult to separate by gel electrophoresis, so their co-amplification is not advised. Note that SiHa DNA cannot be used as a positive control for the E6 amplification since this region is inefficiently amplified from this cell line.

Protocol 6. Amplification of HPV E6

Materials

Same as *Protocol 4*, except PCR primers

HPV E6 primers: WD72, WD76, WD66, WD67, WD154, each at 100 μM (see *Table 2*)

Method

1. Use the PCR preparation guidelines and reaction conditions outlined in *Protocol 4* with the following primer substitutions:
   - WD72 @ 100 μM   0.1 μl
   - WD76 @ 100 μM   0.4 μl
   - WD66 @ 100 μM   0.1 μl
   - WD67 @ 100 μM   0.4 μl
   - WD154 @ 100 μM   0.1 μl
2. Step cycle as in *Protocol 4* for fresh clinical samples or as in *Protocol 5* for PET samples.

---

4. Analysis of PCR products: electrophoresis

4.1 Polyacrylamide gel electrophoresis (PAGE)

The PCR amplification products described above range in size from 240 to 536 bp and are easily resolved on a 5–7% polyacrylamide or 1.5% agarose gel (see Chapters 1 and 4 for details of agarose gel electrophoresis). Polyacryla-

Table 2. PCR primer sequences for the HPV E6 gene

Name	5'→3' nucleotide sequence	Target	Strand
WD72	CGGTCGGGACCGAAAACGG	HPV E6	+
WD76	CGGTTSAACCGAAAMCGG	HPV E6	+
WD66	AGCATGCGGTATACTGTCTC	HPV E6	−
WD67	WGCAWATGGAWWGCYGTCTC	HPV E6	−
WD154	TCCGTGTGGTGTGTCGTCC	HPV E6	−

mide gels provide better resolution and greater sensitivity. Gel data should never be used to determine the presence of HPV, but an assessment of sample sufficiency can be made based on the efficiency of the β-globin amplifications. Samples that are sufficient for analysis yield a visible β-globin amplification product. The inability of a specimen to amplify can be caused by a lack of intact DNA or by the presence of an inhibitor. The absence of primer–dimer amplification indicates inhibition. Polyacrylamide gel electrophoresis (PAGE) is described in *Protocol 7*.

Protocol 7. PAGE analysis of PCR products

Materials
- 7% polyacrylamide gel
- TBE buffer: 50 mM Tris–HCl, 67 mM boric acid, 1 mM EDTA
- Ethidium bromide, 10 mg/ml in sterile distilled water
- 10 × Gel loading dye: 2.5 mg/ml bromophenol blue, 2.5 mg/ml xylene cyanol, 250 mg/ml Ficoll 400, 0.1 M EDTA [pH 8.0], 0.1% SDS
- Molecular weight marker, 50 ng DNA/μl (123 bp ladder, for example)

Method
1. Prepare a 1-mm-thick, 7% bis-acrylamide gel in TBE buffer. Cast sample wells with a 10- or 14-well comb. Gels can be pre-cast, and stored at 4 °C with combs in place for up to a week.
2. Combine 5 μl of PCR product and 5 μl of 2 × gel loading dye and load into the sample well. Include a lane with the molecular weight marker (250–500 ng) on every gel. Run gels until the bromophenol blue dye just leaves the bottom of the gel.
3. Stain the gel in a dilute ethidium bromide solution (10 μg/ml) for 2 min and photograph using an orange filter and UV transilluminator.

4.2 Digestion analysis of the HPV L1 fragment

If the amplification has produced clean and abundant L1 PCR product, aliquots of the PCR can be digested separately with different restriction enzymes to yield a digestion pattern that is unique to each virus type (5). Subtle differences between two patterns (such as a gain or loss of a restriction site) likely reflect a variant of a type rather than a distinct type.

This restriction analysis can be used for typing small numbers of samples. In addition, groups of samples that do not hybridize to any of the type-specific probes can be further classified and characterized by restriction patterns. In our laboratory, we have used this method to distinguish over 40 different restriction patterns, many of which we believe to be novel HPV types. The procedure is given in *Protocol 8*.

Protocol 8. Restriction digestion analysis of L1 fragments

Materials
- *Same as Protocol 7*
- 10 × *Bgl*II buffer: 1 M NaCl, 100 mM MgCl$_2$, 100 mM Tris–HCl [pH 7.5], 100 mM β-mercaptoethanol
- Restriction enzymes at 4–20 U/μl:
    *Bam*H1, *Dde*1, *Hae*III, *Hin*F1, *Pst*1, *Rsa*1, *Sau*3A1
- Amplification reactions (L1 or L1/β-globin) from samples of interest

Method
1. Determine the amount of each amplification reaction that will give a visible L1 product on a gel, then multiply this volume by eight. For the set of eight digestions, mix the PCR DNA with 12 μl 10 × *Bgl*II buffer, and bring up to a final volume of 112 μl with water. Mix well.
2. Separate the digestion mix into eight aliquots of 14 μl each.
3. Add 1 μl of the appropriate restriction enzyme to each of the seven tubes. Keep one tube as the undigested control. Mix well and spin briefly if necessary.
4. Incubate at 37°C for 1 h.
5. Add 1.7 μl 10 × gel loading buffer and load the entire digestion into sample well. Use the electrophoresis guide-lines outlined in *Protocol 7*.

---

5. Analysis of PCR products: dot blot hybridization

Sensitive and specific detection of HPV DNA is accomplished through dot blot hybridization. Samples are applied to nylon membranes and hybridized to either long generic probes, as with L1 products, or oligonucleotide probes, as with E6 products and typing of L1 products. Because of the high homology between viral types in the L1 region, it is possible to detect a broad spectrum of types with a mixture of three long probes specific to HPV 16, 18, and 31 (7). A generic probe has not been designed for the E6 system. However, the type-specific oligonucleotide probes can be used as a mixture.

The HPV L1 system was designed to amplify a region that contains internal sequences that are specific to each HPV, thus allowing subsequent characterization with type-specific oligonucleotide probes. The E6 products can also be differentiated with oligonucleotide probes.

The β-globin probe can be used to verify the identity of the β-globin PCR product, verify sample sufficiency, and determine the integrity of the prepared dot blots. Both the radioactive and non-radioactive detection methods described below have been used effectively.

5.1 Dot blot preparation

The use of the dot blot format for DNA detection allows the rapid preparation and efficient analysis of large numbers of samples. Blots containing 80–90 reactions can be prepared in replicate and hybridized in parallel with different HPV type-specific probes. In addition, these blots can be stripped of probe and re-hybridized allowing for repeat experiments and additional HPV typing without the need to prepare additional membranes. The procedure in *Protocol 9* outlines the steps for preparing dot blots of amplified β-globin and HPV sequences for use in the detection of DNA with labelled probes.

Protocol 9. Preparation of dot blots

Materials
- Denaturation solution: 0.4 M NaOH, 25 mM EDTA
- 20 × SSC: 0.3 M sodium citrate, 3 M NaCl, pH 7.0
- 20 × SSPE: 3.6 M NaCl, 0.2 M Na phosphate, 0.11 M NaOH, 0.02 M EDTA, pH 7.4
- Nylon membrane (for example, Biodyne-B, ICN, Irvine, California, USA)
- Dot blot manifold[a]

Method
1. Cut membranes to the desired size to fit the manifold. Prepare as many replicate blots as will be needed for typing. Pre-wet the membranes in distilled, deionized water for 20 min. Each blot should contain a set of L1 or E6 hybridization controls that include human HPV-negative controls and plasmid amplifications of the different HPV types.
2. Add 3 μl of PCR product plus 100 μl denaturing solution per dot (for

Protocol 9. *Continued*

   example, for five replicate blots, add 15 µl PCR product to 500 µl denaturing solution). Allow the DNA to denature for at least 10 min before loading.

3. Assemble the dot blot manifold and load 100 µl of the denatured sample solution to the appropriate well. Load the samples in the same location on all replicate membranes. Prepare separate blots for L1 and E6 products since they will be hybridized with different probes. When all samples have been added, apply the vacuum.

4. Rinse each well with 400 µl of 20 × SSC, and re-apply the vacuum. Disassemble the manifold and label the blot.

5. Fix the PCR product to the membrane by placing it DNA-side up in a UV-crosslinker (Stratagene, La Jolla, California, USA) and setting the instrument to 50 mJoule/cm. Alternatively, wrap the membranes in plastic and place DNA-side down on a UV transilluminator (used for gel photography) for 3 min.

6. Rinse the membranes in 2 × SSPE and store at 4°C in 2 × SSPE until ready to hybridize.

   a Dot blots can also be prepared by hand (see Chapter 3).

5.2 Generic L1 probe synthesis

The generic L1 probes are synthesized from very dilute solutions of the L1 PCR fragments of HPV 16, 18, and 31. The three resulting PCR fragments used together detect a broad spectrum of HPV types. To synthesize the probes, amplifications are performed separately with nested type-specific primers (see *Table 3*) to each of the MY09/MY11-generated L1 fragments. These primers are designed to target type-specific sequences that lie inside the MY09 and MY11 sequences such that the primer sequences are not included in the probe sequence. (Note: the omission of the HPV 31-specific probe lessens the detection spectrum only slightly. Thus it is possible to use only the HPV 16 and 18 probes with relative success.)

   Either radioactive ^{32}P or biotin can be incorporated into the probe by using labelled dNTPs. The protocols for both labelling methods are described below. Several kits for non-radioactive detection of biotin-labelled probes are currently available.

Table 3. PCR primers for the synthesis of the generic L1 probe

Name	5'→3' nucleotide sequence	HPV type	Strand
MY74	CATTTGTTGGGGTAACCAAC	Type 16	+
MY75	TAGGTCTGCAGAAAACTTTTC	"	−
MY76	TGTTTGCTGGCATAATCAAT	Type 18	+
MY77	TAAGTCTAAAGAAAACTTTTC	"	−
MY49	TATTTGTTGGGGCAATCAG	Type 31	+
MY50	CTAAATCTGCAGAAAACTTTT	"	−

Protocol 10. Synthesis of the labelled generic L1 probe

A. Radioactive

Materials

- Same as *Protocol 4*, except PCR primers
- Type-specific nested generic L1 probe primers
  MY74, MY75, MY76, MY77, MY49, and MY50, each at $100 \mu M$
  (see *Table 3*).
- HPV L1 PCR products from HPV 16, 18, and 31 (diluted 1:1000)
- α-^{32}P-labelled dATP, dCTP, dGTP, and dTTP at 800 Ci/mmol (NEN
  Dupont, Wilmington, Delaware, USA)
- Spin column
- Salmon sperm DNA

Method

1. Since the starting material is PCR product it is important not to use the
   PCR-clean area for preparing the probe synthesis reactions.
2. Set up six separate reactions so that one amplification of each of the three
   types can be radioactive and one non-radioactive. Be sure that the appropriate primers are mixed with the corresponding target (see *Table 3*). Use the
   same reaction and thermocycling profile described in *Protocol 4*, but with
   $50 \mu M$ each unlabelled dNTP, $0.4 \mu M$ of each primer, and 1–2 μl of the
   appropriate diluted L1 product in a final volume of $50 \mu l$. Add 62.5 pmol
   ($50 \mu Ci$) of each α-^{32}P-labelled dNTP to the radioactive reactions.
3. Omit the mineral oil overlay and run 25 temperature cycles as in *Protocol
   4* except with denaturation and annealing times of 30 sec.
4. To assess the efficiency of the amplification, visualize the PCR products
   from the identical concurrent unlabelled reactions using electrophoresis
   and ethidium bromide staining (see *Protocol 7*).
5. Purify the radiolabelled 400 bp fragments using a spin column (for example,
   G50 Sephadex: see Chapter 2).
6. Before hybridization, denature the three fragments at 95 °C for 10 min in
   the presence of 2 mg/ml sheared salmon sperm DNA (see *Protocol 11*).
7. Hybridize the membranes with a mixture of 10^5 c.p.m. of each probe per
   millilitre of hybridization buffer.

B. Non-radioactive (biotinylated)

Materials

- Same as *Protocol 4*, except PCR primers

Protocol 10. *Continued*

- Type-specific nested generic L1 probe primers
  MY74, MY75, MY76, MY77, MY49, and MY50 each at 100 μM (see *Table 3*)
- HPV L1 PCR products from HPV 16, 18, and 31 (diluted 1:1000)
- Bio-11-dUTP (Sigma, St. Louis, Missouri, USA)
- 5 M Ammonium acetate
- Isopropanol

Methods

1. PCR reactions are similar to those described in *Protocol 4*. However the primer, magnesium, and nucleotide concentrations are altered.
2. Each 100 μl reaction contains 1–2μl of the appropriate diluted L1 product, 50 mM KCl, 10 mM Tris–HCl [pH 8.5], 2.5 U *Taq* DNA polymerase, 200 μM dATP, dCTP, and dGTP, 100 μM dTTP, and 100 μM Bio-11-dUTP.
3. Set up three reactions with the following changes in primer and magnesium concentrations.
   - HPV16 reaction: 0.1 μM (10 pmol) each primer (MY74, MY75), 4 mM MgCl$_2$
   - HPV18 reaction: 0.5 μM (50 pmol) each primer (MY76, MY77), 8 mM MgCl$_2$
   - HPV31 reaction: 0.1 μM (10 pmol) each primer (MY49, MY50), 6 mM MgCl$_2$

   Separate 10 × PCR buffers can be made for each reaction, or MgCl$_2$ can be added separately.
4. Add 100 μl mineral oil and run 25 temperature cycles as in *Protocol 4*, except with denaturation and annealing times of 30 sec.
5. To assess the efficiency of the amplification, visualize the PCR products, using electrophoresis and ethidium bromide staining (see *Protocol 7*). Biotinylated DNA fragments have slightly retarded migration.
6. Purify the biotinylated fragments by isopropanol precipitation by adding 0.4 vol. 5 M ammonium acetate and 2 vol. isopropanol (13). Resuspend the fragments in TE buffer.
7. The optimum ratio and concentration of probe fragments to be used in the hybridization must be determined by hybridizing test strips with known amounts of HPV DNA. High probe concentrations will produce undesirable background signal, whereas lower concentrations will decrease the sensitivity. Both the sensitivity and signal-to-noise ratio are established subjectively.

146

8. Before hybridization, denature the probe mixture at 95°C for 10 min in the presence of 2 mg/ml sheared salmon sperm DNA (see *Protocol 11*).

5.3 Hybridization and detection

Once the PCR products are immobilized on membranes and the probes labelled, the blots can be hybridized to determine the type and presence of HPV DNA. *Table 4* contains the sequences for oligonucleotide probes specific for HPV L1 type 6, 11, 16, 18, 31, 33, 35, 39, 45, 51, and 52; and E6 type 6, 11, 16, 18, 31, 33, and 45; and β-globin. Probes for the same type are used together. In addition, probes for different types can be used as a mixture to allow grouping of types. For example, the probes for types 6 and 11 are usually used together. As the sequences of more HPV types become available, this list of type-specific probes will be expanded.

These oligonucleotide probes can be radioactively labelled using a kinase reaction with γ-^{32}P ATP (13 and see Chapter 2). A standard labelling reaction yields a specific activity of $0.5–1.0 \times 10^7$ c.p.m./pmol probe. Alternatively, the probes can be biotinylated either by being synthesized with a terminal biotin or by being end-labelled using a terminal deoxynucleotide transferase (TdT) reaction with biotin–dUTP (14 and see Chapter 2).

Since the different HPV types have a high degree of homology in the L1 region, these probes tend to cross-react with other types when the stringency of hybridization is lowered. Thus it is very important to accurately measure and maintain the wash temperatures. A few degrees in either direction will seriously affect sensitivity and specificity. The L1 system requires a mixture of two oligonucleotide probes for each HPV type so as not to miss variants that vary in nucleotide sequence in one of the probe regions.

The Amersham Enhanced ChemiLuminescence (ECL) has been utilized for the non-radioactive detection of HPV and β-globin PCR products. One drawback is that any peroxidases or pseudo-peroxidases in the sample will react with the kit reagents and create a false-positive signal. When using the crude sample preparation methods described here, there may be difficulty with bloody samples. In addition, the ECL system requires the use of a particular membrane, Biodyne B (ICN, Irvine, California, USA).

The hybridization and detection of radioactive and non-radioactive probes are detailed in *Protocol 11*.

Protocol 11. Hybridization and detection

Materials
- Pre-treatment buffer: 0.1 × SSPE, 0.5% SDS
- Generic probe hybridization buffer: 5 × SSPE, 0.1% SDS, 100 µg/ml single-stranded sheared salmon sperm DNA

Protocol 11. *Continued*

- Oligonucleotide probe hybridization buffer: 5 × SSPE, 0.1% SDS
- Blot wash solution: 2 × SSPE, 0.1% SDS
- Labelled DNA probes (see *Protocol 10* and *Table 4*):
    Generic L1 probe mix (HPV 16, 18, 31)
    Type-specific L1 oligonucleotide probes
    Type-specific E6 oligonucleotide probes
    β-globin oligonucleotide probe
- For Enhanced ChemiLuminescent detection:
    Streptavidin–horse radish peroxidase (SA–HRP) conjugate, 1 mg/ml
    (Vector Laboratories, Burlingame, California, USA)
    ECL reagents 1 and 2 (Amersham, Arlington Heights, Illinois, USA)

Method

1. Add membranes to a large volume of pre-heated pre-treatment buffer. Incubate with shaking for 30 min at 65°C.

2. After pre-treatment, place up to two membranes (DNA side out) into a hybridization bag. Add hybridization buffer (without probe) to the bag. Use 20 ml buffer for two 100 cm^2 blots or 10 ml for one blot. Pre-hybridize at 65°C for 15 min. For the generic L1 probe, use the hybridization buffer containing salmon sperm DNA.

3. For the L1 generic probe hybridizations, the probe mixture must be denatured in the presence of 2 mg/ml salmon sperm DNA before use. Add the amount needed (see *Protocol 10*) of generic L1 probe mix to a 1.5-ml microfuge tube and heat for 10 min at 95°C. The probes must either be added immediately to the appropriate volume of pre-heated hybridization buffer or be rapidly cooled in a dry ice/alcohol bath. Slow cooling will cause re-annealing of the DNA strands which will result in decreased sensitivity of the hybridization.

4. For the oligonucleotide probes, prepare separate hybridization solutions for each HPV type or type grouping. The ^{32}P-labelled probes should be used at a final concentration of 10^5 c.p.m. of each probe/ml. The biotinylated probes should be used at a final concentration of 0.5 pmol each probe/ml.

5. Drain the hybridization solution from the bags. Add 10–20 ml of hybridization buffer containing probe to each bag. Remove large air-bubbles, seal the bag, and mix the solution thoroughly. Hybridize at 55°C for at least 1.5 h.[a] Massage and turn over the bags after 45 min.

6. Remove the membranes from the bags and briefly rinse in pre-warmed wash solution (37–45°C). Place the membranes into a wash dish containing a large volume of wash solution pre-heated to 56°C. Wash the membranes at 56°C in a shaking water bath for 10 min. Drain, add fresh

pre-heated wash solution, and wash for an additional 10 min. Some of the E6 oligonucleotide probes require different wash temperatures: 45°C for WD170 and WD171; 52°C for RR1 and RR2.

7. For radioactively labelled probes, wrap the blots in plastic and expose to X-ray film overnight (see Chapter 1). For biotinylated probes, proceed to streptavidin binding and ECL detection (steps 8–11).[b]

A. ECL detection

8. Using a plastic tray, add SA-HRP conjugate to 250 ml of wash solution to make a final concentration of 30 ng/ml of SA-HRP for the generic probe mix or 40 ng/ml for the oligonucleotide probes. Mix gently. Transfer the blots to the SA-HRP solution. The blots should be completely submerged. Allow binding to occur with gentle agitation for 15 min at room temperature.

9. Drain the SA-HRP solution, transfer the blots to a large plastic container and wash in a large volume of blot wash with vigorous mixing for 10 min. Drain and repeat the wash.

10. Proceed to ECL development outlined in Amersham manufacturer's instructions.

11. Expose to X-ray film at room temperature for 2 h and develop the film. Blots hybridized with the β-globin probe require only one short exposure of 5 min.[b]

---

[a] E6 probes WD170 and WD171 require a 45°C hybridization temperature.
[b] The blots probed with the generic probe fragment cannot be re-used. However, blots probed with the type-specific oligonucleotide probes can be re-used following the removal of the probe either by standard alkali or boiling treatment (13). Stripped blots can be stored in 2 × SSPE at 4°C.

---

6. Applications to laboratory research

The clinical utility of these sensitive and broad-spectrum HPV detection methods is yet to be determined. The relationship between the presence of HPV and the development of disease must be understood before these methods can be used as a clinical diagnostic tool. We are concerned that the premature use of this method clinically to screen for disease may cause confusion and possibly inappropriate treatment. We foresee an eventual staged introduction of clinical applications, if data from case-control and cohort studies suggest a predictive value. On the other hand, this method has numerous immediate applications to laboratory research.

These techniques can be employed in molecular epidemiology to determine the prevalence of infection and the epidemiologic risk factors in many different populations. We have shown that the use of this method reduces misclassifica-

Table 4. Oligonucleotide probes for HPV L1, E6, and β-globin amplification products[a]

Name	5′→3′ nucleotide sequence	HPV type
HPV L1 probes		
MY12	CATCCGTAACTACATCTTCCA	6
MY13	TCTGTGTCTAAATCTGCTACA	11
MY125	ACAATGAATCCYTCTGTTTTGG	6 and 11
MY95	GATATGGCAGCACATAATGAC	16
MY133	GTAACATCCCAGGCAATTG	16
WD74	GGATGCTGCACCGGCTGA	18
MY130	GGGCAATATGATGCTACCAAT	18
WD128	TTGCAAACAGTGATACTACATT	31
MY92	CCAAAAGCCYAAGGAAGATC	31
MY16	CACACAAGTAACTAGTGACAG	33
MY59	AAAAACAGTACCTCCAAAGGA	33
MY115	CTGCTGTGTCTTCTAGTGACAG	35
MY117	ATCATCTTTAGGTTTTGGTGC	35
MY89	TAGAGTCTTCCATACCTTCTAC	39
MY90	CTGTAGCTCCTCCACCATCT	39
MY98	GCACAGGATTTTGTGTAGAGG	45
MY69	ATACTACACCTCCAGAAAAGC	45
MY87	TATTAGCACTGCCACTGCTG	51
MY88	CCCAACATTTACTCCAAGTAAC	51
MY81	CACTTCTACTGCTATAACTTGT	52
MY82	ACACACCACCTAAAGGAAAGG	52
HPV E6 probes		
WD133	ACACCTAAAGGTCCTGTTTC	6
WD134	ACACTCTGCAAATTCAGTGC	11
WD103	CAACAGTTACTGCGACG	16
WD170	GCAAGACATAGAAATAA	18
WD132	GACAGTATTGGAACTTACAG	18
WD165	AAATCCTGCAGAAAGACCTC	31
WD166	CCTACAGACGCCATGTTCA	31
RR1	GTACTGCACGACTATGT	33
RR2	ACCTTTGCAACGATCTG	33
WD171	ACAAGACGTATCTATTG	45
β-globin probe		
PCO3	ACACAACTGTGTTCACTAGC	

[a] Degenerate code Y = C + T

tion of HPV status so that it is possible to assess definitively the epidemiology of infection (15). The relationship of HPV to various clinical manifestations, progression and regression of disease, and type-specific risk of disease can also be assessed. These studies are likely to be useful in determining diagnostic utility. The sensitivity of this method allows medical researchers to compare different sampling protocols to determine which are optimal in detecting clinically relevant HPV infection.

Although this method has been used primarily for the study of cervical disease, it can also be utilized to determine the presence or absence of HPV in other diseased tissues. For example, inflammations and cancers of the oral and urogenital region that have no known causal agent can be analysed for the presence of HPV. Anogenital neoplasia as well as HPV-related problems in immunocompromised individuals can also be studied. Overall, the sensitivity, specificity, and broad type-spectrum of this assay provide a reliable means of assessing the presence of HPV. This method is likely to play a key role in our understanding of the etiology and natural history of HPV infection. The applications to laboratory research are limited only by the researchers' imagination.

Acknowledgements

We are indebted to J. Sninsky and S. Williams for their support and encouragement; the hard working members of the HPV laboratory: R. Resnick, Y. Ting, D. Wright, A. Lewis, and G. Eichinger; the DNA synthesis group at Cetus: C. Levenson, D. Spasic, L. Goda, and O. Budker, and R. Kurka for help in manuscript preparation.

References

1. Koutsky, L A., Galloway, D. A., and Holmes, K. K. (1988). *Epidemiol. Rev.,* **10,** 122.
2. Roman, A. and Fife, K. H. (1989). *Clin. Microbiol. Rev.,* **2,** 166.
3. de Villiers, E-M. (1989). *J. Virol.,* **63,** 4898.
4. Manos, M. M., Ting, Y., Wright, D. K., Lewis, A. J., Broker, T. R., and Wolinsky, S. M. (1989). In *Molecular diagnostics of human cancer: cancer cells* (ed. M. Furth and M. Greaves), Vol. 7, pp. 209–14. Cold Spring Harbor Press, Cold Spring Harbor, NY.
5. Ting, Y. and Manos, M. M. (1990). In *PCR protocols: a guide to methods and applications* (ed. M. Innis, D. Gelfand, J. Sninsky, and T. White), pp. 356–67. Academic Press, San Diego, California.
6. Resnick, R. M., Cornelissen, M. T. E., Wright, D. K., Eichinger, G. H., Fox, H. S., ter Schegget, J., and Manos, M. M. (1990). *J. Natl. Cancer Inst.,* **82,** 1477.
7. Bauer, H. M., Ting, Y., Greer, C. E., Chambers, J. C., Tashiro, C. J., Chimera, J., Reingold, A., and Manos, M. M. (1991). *J. Am. Med. Assoc.,* **265,** 472.
8. Greer, C. E., Peterson, S. L., Kiviat, N., and Manos, M. M. (1991). *Am. J. Clin. Pathol.* **95,** 117.
9. Wright, D. K. and Manos, M. M. (1990). In *PCR protocols: a guide to methods and applications* (ed. M. Innis, D. Gelfand, J. Sninsky, and T. White), pp. 153–8. Academic Press, San Diego, California.

151

10. Mullis, K. B. and Faloona, F. A. (1987). *Methods in enzymology*, Vol. 155 (ed. R. Wu), pp. 335–50. Academic Press, New York.

11. Saiki, R. K., Scharf, S., Faloona, F., Mullis, K. B., Horn, G. T., Erlich, and H. E., Arnheim, N. (1985). *Science,* **230,** 1350.

12. Kwok, S and Higuchi, R. (1989). *Nature,* **339,** 237.

13. Maniatis, T., Fritsch, E. F., and Sambrook, J. (ed.) (1982). *Molecular cloning, a laboratory manual.* Cold Spring Harbor Press, Cold Spring Harbor, NY.

14. Cook, A. F., Vuocolo, E., and Brakel, C. L. (1988). *Nucleic Acids Res.,* **16,** 4007.

15. Ley, C., Bauer, H. M., Reingold, A., Schiffman, M. H., Chambers, J. C., Tashiro, C. J., and Manos, M. M. (1991). *J. Natl. Cancer Inst.,* **83,** 997.

7

The polymerase chain reaction for human papillomavirus screening in diagnostic cytopathology of the cervix

JAN M. M. WALBOOMERS, PETER W. J. MELKERT,
ADRIAAN J. C. VAN DEN BRULE,
PETER J. F. SNIJDERS, and CHRIS J. L. M. MEIJER

1. Introduction

Cytomorphological examination of cervical smears is the most widely applied and accepted screening method for cervical cancer, originally introduced by Papanicolau. The value of the Pap screening technique is limited by several factors—for example, sampling errors by the clinician and screening errors by the microscopist. Thus, there is a need for a more accurate and faster screening technique. Cytomorphological abnormalities in cervical smears may be classified according to the Pap classification. Pap I and II correspond to no significant morphological changes; Pap IIIa to mild to moderate dysplasia; Pap IIIb to severe dysplasia; Pap IV to carcinoma *in situ*; and Pap V to invasive carcinoma. Several studies indicate that only a small proportion (10–30%) of women with Pap III or IV smears will eventually progress to invasive carcinoma (1–3). Since all patients with cytomorphological abnormalities of the cervix are treated at present, this results in a considerable over-treatment of patients. Thus, additional criteria are needed to predict more accurately the clinical outcome of cervical lesions in individual women.

Recent developments in tumour virology have shown that specific human papillomaviruses (HPVs) play an important part in the development of cervical cancer (4, 5). In general, HPV 6 and 11 have been associated with benign cervical lesions and are referred to as non-oncogenic or low-risk types. HPV 16, 18 and to a lesser extent HPV 31, 33, 35 have been found mainly in mild and severe dysplasia and cervical carcinoma and are considered as oncogenic or high-risk types (6–12).

To date, 60 different HPV genotypes have been isolated of which 24 are associated wth the genital mucosa (13). Because HPV cannot be cultured *in vitro* and suitable HPV type-specific antibodies are not available for immuno-histochemistry, HPV genotypes in cervical smears must be identified by nucleic acid detection. The most sensitive HPV DNA detection method at the time of writing is the polymerase chain reaction (PCR) (14, 15), which is theoretically able to detect one copy of a target sequence in a given sample. To extend our knowledge about the role of HPV in the development of cervical cancer and to evaluate the diagnostic value of HPV detection in cervical smears in tracing pre-malignant and malignant cervical lesions, a sensitive HPV PCR method was developed and applied by our group to large numbers of cervical scrapes with normal and abnormal cytology (16–20). In this chapter, we will describe:

(a) Sample preparation for PCR

(b) HPV detection in cervical scrapes

(c) Clinical aspects: the possible implications for cervical cancer screening

2. Sample preparation for PCR

Cervical scrapes were derived from the transformation zone of the cervix uteri. The use of the Cervex brush is recommended because of the excellent quality of the smear for microscopical examination (Dr O. P. Bleker, personal communication). The first smear is made for routine cytological examination. For HPV detection, the remaining material from the first brush and an additional scrape are put in 5 ml sterile phosphate-buffered saline (PBS) with 0.05% merthiolate. Details of sample preparation are given in *Protocol 1*. The most important prerequisite to the use of PCR for large-scale screening programmes is the omission of laborious DNA extraction procedures. Previous studies have already reported the performance of PCR directly on single cells (21) or cell suspensions (22). No reproducible results were obtained when cells were subjected to either proteinase K incubation (21, 23) or alkaline denaturation followed by neutralization (24). Pretreatment by freeze-thawing before adding the PCR mixture resulted in consistently successful amplification of the HPV target sequences. The efficiency of amplification after this pretreatment was similar to the PCR performed on purified DNA as studied with cervical scrapes which were previously found to contain HPV 16, 18, and 33 (19).

Protocol 1. Sample preparations for PCR[a]

Materials
- PBS: 10 mM phosphate, 150 mM NaCl, pH 7.2
- Cervex brushes (International Medical Products, Zutphen, The Netherlands)
- 10 mM Tris–HCl, pH 8.1

Method

1. Collect scrapes in 5 ml of sterile PBS containing 0.05% mertiolate and vortex vigorously.

2. Remove Cervex brushes with sterile forceps (intermittently heated).

3. Pellet the cells by centrifugation at 3000 g for 10 min.

4. Re-suspend the cells in 1 ml of 10 mM Tris–HCl pH 8.1 with a disposable pipette (transfer pipettes, Sarstedt).

5. Take four 10 μl aliquots for PCR purposes[b] and store all aliquots and remaining samples at −70°C.

6. Thaw one 10 μl aliquot for each PCR assay.

7. Denature the aliquot for 5 min at 100°C, then cool on ice and centrifuge for 1 min at 3000 g.

8. Add the desired PCR mixture (see *Protocol 2*).

[a] The whole procedure except centrifugation should be carried out in a laminar flow hood.
[b] Reaction tubes containing screw caps (Sarstedt) are recommended. Furthermore, positive displacement pipettes (Microman, Gibson, France) should be used.

3. Detection of human papillomavirus in cervical scrapes

3.1 General and specific primers for HPV

Because PCR is the most sensitive method for the detection of HPV DNA in cervical scrapes (17, 24), a study was initiated to develop a new rapid and reliable HPV screening method based on PCR. It was necessary to develop a single PCR assay for the detection of multiple HPV types simultaneously. For this reason, primer sequences were selected that are conserved among a broad spectrum of HPV types. Matrix comparison (25) of the sequenced types HPV 6b, 11, 16, 18, 31, and 33 revealed that the most conserved regions are located within the E1 and L1 open reading frames (ORFs) (26). Two 20 bp sequences within the L1 ORF were found that are highly conserved among the sequenced HPV types and could therefore be used as general primers for the PCR (GP 5/6, see *Table 1* and ref. 16). These general primers possess mismatches with the different HPV targets making low stringency conditions of primer annealing necessary for the detection of a broad spectrum of HPV genotypes. The optimization of the GP-PCR is described in detail by Snijders *et al.* (1990) (16). The sequences recognized by these primers span a region of approximately 140 to 150 bp in all examined HPV genomes. Because only 6 of the genital HPV types had been sequenced at the time of primer selection, the general properties of the selected primer set

Table 1. General primers for HPV detection specific for genital HPV genotypes:

GP 5	GP 5: 5' TTTGTTACTGTGGTAGATAC 3'	GP 6: 3' ACTAA AT GTC AAATAA AAAG 5'	Mismatches GP 5	Mismatches GP 6
HPV 6b	. . . .	. . . . T	0	1
HPV 11	. . . .	. . . .	0	0
HPV 16	. . . . T . . T	. . . .	2	0
HPV 18	. . . .	. . . . . . C	0	1
HPV 31	. . . .	. . . . T . . . . T	0	2
HPV 33	. . . .	. . . G . . . . C	0	2

Alignment of primer GP5 and GP6 sequences with the corresponding sequences within the L1 ORF of the sequenced genital HPV types. Dots represent identical bases whereas mismatched bases are indicated. The number of mismatches is given on the right. The nucleotide positions of the first nucleotide of the GP5/6 matched sequences are as follows: HPV 6b: 6764/6883; HPV 11: 6749/6868; HPV 16: 6624/6746; HPV 18: 6600/6725; HPV 31: 6542/6664; HPV 33: 6581/6700. Locations are numbered according to the published sequence data: HPV 6b (31); HPV 11 (32); HPV 16 (33); HPV 18 (34); HPV 31 (35); HPV 33 (36).

could only be determined by subjecting a large group of cloned HPV genotypes to PCR. *Figure 1A* shows that, in addition to sequenced HPV types 6b,
11, 16, 18, 31, and 33, the GP 5/6 primer set yielded 140–150 bp products with
the unsequenced HPV types 13, 30, 32, 39, and 45. The GP 5/6 oligomers did
not direct for the amplification of pBR 322 DNA, indicating that the PCR
products were HPV-specific and were not the result of cross-hybridization
with vector sequences. This was also confirmed after blotting and hybridization under low stringency using a probe mixture consisting of amplification
products specific for HPV 6, 11, 16, 18, 31 and 33: all 140–150 bp fragments
could be detected (see *Figure 1B*). Specificity of the PCR products was also
shown by dot blot analyses and RsaI restriction enzyme digestion as previously
described (16) and by sequence analysis. Furthermore, it was demonstrated
empirically that up to three mismatches with one or both primers could be
detected. With respect to sensitivity, 0.1 to 1 fg of pHPV DNA that shows up

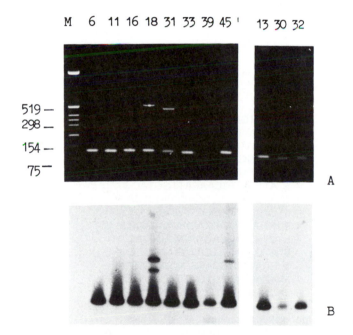

Figure 1.
A. PCR of 1 ng pHPV of several HPV types using the GP 5/6 primer pair under moderate
stringency conditions in the presence of 3.5 mM Mg²⁺. PCR products are shown after
electrophoresis, ethidium bromide staining, and UV irradiation, M, pBR322 fragments
(BP) digested with *Hinf I*.
B. Hybridization after blotting under low stringency conditions (Tm-33) with ³²P-labelled
GP-PCR amplified products of HPVs 6, 11, 16, 18, 31, and 33.

to three mismatches with one or both primers could be detected. This corresponds to approximately 7 to 70 viral genomes (16).

It was also demonstrated that the GP 5/6 primer pair allowed amplification of different HPV targets in cervical scrapes. However, many co-amplified cellular DNA fragments were observed in PCR products. These co-amplified products made it necessary to confirm the presence of HPV by hybridization. Type-specific primers for HPV 6, 11, 16, 18, 31, and 33 were selected (see *Table 2*), since the nucleotide sequence of these genotypes is known. These primers were chosen in such a manner that it was possible to discriminate between several HPV types by the different lengths of the PCR products generated (17).

3.2 HPV screening strategy

With the development of the direct PCR, which omits laborious DNA extraction, a new screening strategy can be introduced (19). This is outlined schematically in *Figure 2*. Using general primers (GP 5/6), cervical scrapes

---

Table 2. Specification of type-specific primers and probes

Type-specific primers:		Amplimer length (base-pairs)
HPV 6.1:	+ 5′ TAGTGGGCCTATGGCTCGTC 3′	
HPV 6.2:	− 5′ TCCATTAGCCTCCACGGGTG 3′	280
HPV 11.1:	+ 5′ GGAATACATGCGCCATGTGG 3′	
HPV 11.2:	− 5′ CGAGCAGACGTCCGTCCTCG 3′	360
HPV 16.1:	+ 5′ TGCTAGTGCTTATGCAGCAA 3′	
HPV 16.2:	− 5′ ATTTACTGCAACATTGGTAC 3′	152
HPV 18.1:	+ 5′ AAGGATGCTGCACCGGCTGA 3′	
HPV 18.2:	− 5′ CACGCACACGCTTGGCAGGT 3′	216
HPV 31.1:	+ 5′ ATGGTGATGTACACAACACC 3′	
HPV 31.2:	− 5′ GTAGTTGCAGGACAACTGAC 3′	514
HPV 33.1:	+ 5′ ATGATAGATGATGTAACGCC 3′	
HPV 33.2:	− 5′ GCACACTCCATGCGTATCAG 3′	455
PCO3[a]:	+ 5′ ACACAACTGTGTTCACTAGC 3′	
PCO4[a]:	− 5′ CAACTTCATCCACGTTCACC 3′	102

Type-specific oligonucleotide probes:
HPV 6:	+ 5′ CATTAACGCAGGGGCGCCTGAAATTGTGCC 3′
HPV 11:	+ 5′ CGCCTCCACCAAATGGTACACTGGAGGATA 3′
HPV 16:	+ 5′ GCAAACCACCTATAGGGGAACACTGGGGCA 3′
HPV 18:	+ 5′ TGGTTCAGGCTGGATTGCGTCGCAAGCCCA 3′
HPV 31:	+ 5′ ACCTGCGCCTTGGGCACCAGTGAAGGTGTG 3′
HPV 33:	+ 5′ CAAATGCAGGCACAGACTCTAGATGGCCAT 3′

[a] PCO3 and PCO4 are β-globin primers as described by Saiki *et al.* (1985) (6).

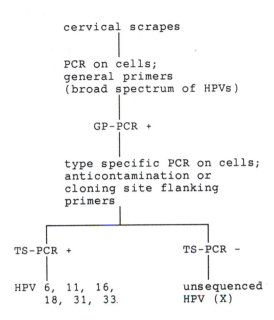

Figure 2. Screening strategy for HPV detection in cervical scrapes by direct GP/TS-PCR.

can be pre-screened for the presence of a broad spectrum of HPV genotypes, if not all genital HPV types, in a single reaction. After pre-screening by GP-PCR, the sequenced HPV types can be identified by HPV type-specific PCR (TS-PCR). The results of the application of the GP-PCR (see *Protocol 2*) using GP 5/6 primers on crude cell extracts of 40 cervical scrapes are shown in *Figure 3*. In *Figure 3A*, the PCR products are shown after agarose gel electrophoresis. The presence of many bands is obvious in most cases. After hybridization (see *Protocol 4*) under low stringency conditions, which allow the cocktail probe (see *Protocol 5*) to hybridize with a broad spectrum of HPV types, strongly and weakly positive bands were observed (see *Figure 3B*). These samples were subjected to HPV type-specific PCR for the detection of sequenced HPV types.

Protocol 2. General primer mediated PCR (GP-PCR) on cervical smears

Materials

- 10 × GP-PCR solution: 500 mM KCl, 100 mM Tris–HCl pH 8.3, 35 mM $MgCl_2$, 0.1% gelatin
- dNTP solution: 2 mM each of dATP, dGTP, dCTP, and dTTP

Protocol 2. *Continued*

- AmpliTaq DNA polymerase (5 U/μl) (Perkin–Elmer Cetus, USA)
- GP5 and GP6 primers (see *Table 1*): 100 pmol/μl

Method

1. Prepare a PCR mixture for 100 reactions by adding the following components in this order on ice:

   - Distilled water   2930 μl
   - 10 × GP-PCR solution   500 μl
   - dNTP solution   500 μl
   - GP5   25 μl
   - GP6   25 μl
   - AmpliTaq DNA polymerase   20 μl

2. Add 40 μl of PCR mixture to 10 μl of crude cell extract (see *Protocol 1*).

3. Add several drops (approx. 100 μl) of mineral oil.

4. Perform 40 cycles of PCR amplification using a Biomed Thermocycler (Amtelstad, The Netherlands) (or equivalent) after a 4 min denaturation step at 95 °C. Each cycle should include a denaturation step of 95 °C for 1 min, an annealing step of 40 °C for 2 min and an elongation step of 72 °C for 1.5 min. The final elongation step should be prolonged for a further 4 min.

5. Subject 10 μl of the PCR product to 1.5% agarose electrophoresis in Tris–borate buffer and ethidium bromide (see Chapters 1 and 4).

The results of HPV TS-PCR (see *Protocol 3*) of the selected GP-PCR samples are shown in *Figure 4*. Samples were tested by using a mixture of primer sets specific for HPV 6, 16, 33, and HPV 11, 18, and 31. After Southern blotting and hybridization (see *Protocol 4*) with HPV 6, 16, 33, and HPV 11, 18, and 31 specific oligonucleotide probes (see *Protocol 6*), the HPV type can be scored on the basis of hybridization signal and length of the amplified fragment. Of the 14 GP-PCR positive scrapes, 7 samples appeared positive by TS-PCR containing HPV 16 in 4 cases, and HPV 18, HPV 31, and HPV 6 in one case each. All these scrapes gave a strong hybridization signal in the GP-PCR. Four samples with a strong GP-PCR hybridization signal appeared to be negative by TS-PCR, indicating the presence of unsequenced known or new HPV types, and marked as HPV X. Unsequenced HPV types can be identified by dot-blot hybridization using known cloned unsequenced HPVs as probes (18). In many cases, HPV 13, 30, 31, 45, and 51 specific DNA could be identified (18). The possible new HPV types can be subjected to sequence analysis (see Chapter 5) and molecular cloning.

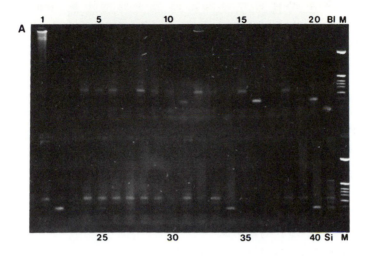

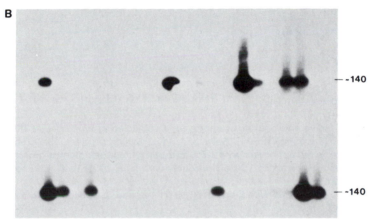

Figure 3. Detection of HPV genotypes in cervical scrapes by direct PCR.
A. Crude cell suspensions of scrapes 1–40 were tested by general primer mediated PCR (GP-PCR). In addition a positive control (Si; 10^3 SiHa cells which contain 1–10 copies of HPV 16 per genome) and a negative control (B1; distilled water) were analysed. GP-PCR products are shown after electrophoresis on a 1.5% agarose gel and ethidium bromide staining.
B. Southern blot analysis of GP-PCR products under low stringency hybridization conditions with labelled GP-PCR products derived from cloned sequenced HPV types 6, 11, 16, 18, 31, and 33. Washes were also performed under conditions of low stringency.

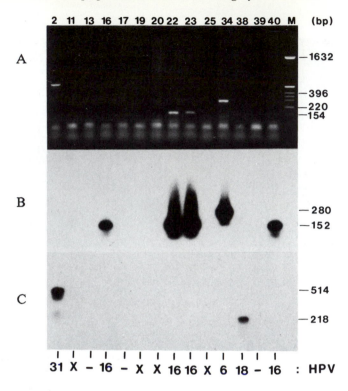

Figure 4.
A. GP-PCR positive and doubtful scrapes were subjected to type specific PCR (TS-PCR) using a mixture of HPVs 6, 11, 16, 18, 31, and 33 specific primers. TS-PCR products are shown after electrophoresis on a 1.5% agarose gel and ethidium bromide staining.
B. Southern blot analysis of TS-PCR products with labelled HPVs 6, 16, and 33 specific oligonucleotides.
C. Southern blot analysis with labelled HPVs 11, 18, and 31 specific oligonucleotides. M: pBR 322 × *Hinf I*.

Protocol 3. Type-specific PCR (TS-PCR) on cervical smears

Materials

- 10 × TS-PCR solution: 500 mM KCl, 100 mM Tris–HCl pH 8.3, 15 mM MgCl$_2$, 0.1% gelatin
- dNTP solution: 2 mM each of dATP, dGTP, dCTP, and dTTP
- AmpliTaq DNA polymerase (5 U/μl) (Perkin–Elmer, Cetus, USA)
- Primers for HPV 6, 11, 16, 18, 31, and 33, each at 100 pmol/μl (see *Table 2*)

162

Method

Two different multiplex PCR assays are performed containing primer pairs specific for HPV 6, 16, 33 and HPV 11, 18, 31 respectively (see *Table 2*).

1. Prepare two PCR mixtures, both suitable for 100 PCR reactions as follows:

    Mixture for HPV 6, 16, 33:

    - Distilled water   2830 μl
    - 10 × TS-PCR   500 μl
    - dNTP solution   500 μl
    - Each of primers HPV 6.1, 6.2, 16.1, 16.2, 33.1, 33.2   25 μl
    - AmpliTaq DNA polymerase   20 μl

    Mixture for HPV 11, 18, 33:

    This is identical to the mixture for HPV 6, 16, 33 except that primers specific for HPV 11, 18, and 33 are used.

2. Add 40 μl of PCR mixture to 10 μl of crude cell extract (see *Protocol 1*).

3. Add several drops (approx. 100 μl) of mineral oil.

4. Perform 40 cycles of PCR amplification using a Biomed Thermocycler (Amtelstad, The Netherlands) (or equivalent) after a 4-min denaturation step at 95 °C. Each cycle should include a denaturation step of 95 °C for 1 min, an annealing step of 55 °C for 2 min and an elongation step of 72 °C for 1.5 min. The final elongation step should be prolonged for a further 4 min.

5. Subject 10 μl of the PCR product to 1.5% agarose electrophoresis in Tris–borate buffer and ethidium bromide (see Chapters 1 and 4).

Protocol 4. Blotting and hybridization

The blotting and hybridization procedure described below can be successfully applied to BioTrace RP (Gelman Science) and GeneScreen Plus (NEN DuPont) charged nylon membranes.

Materials

- Pre-hydridization mixture: for a 1-litre solution, add 248 ml distilled water, 500 ml 0.5 M Na_2HPO_4 adjusted to pH 7.4 with phosphoric acid, 350 ml 20% SDS and 2 ml 0.5 M EDTA, pH 8.0
- Transfer buffer: 0.6 M NaCl/0.4 M NaOH

Protocol 4. *Continued*

- 20 × SSC: 3 M NaCl, 0.3 M sodium citrate, pH 7.0
- 2 × SSC, 0.2 M Tris–HCl, pH 7.5
- Labelled probe (see *Protocols 5* and *6*)
- 3 × SSC, 0.5% SDS

Method

A. Blotting

No pre-treatment of the gel is required for this blotting procedure

1. Make a normal gel set-up for blotting using 0.6 M NaCl/0.4 M NaOH as transfer buffer (see *Figure 1*, Chapter 3).
2. Pre-wet the membrane for 10 min in distilled water and place the membrane on the gel.
3. Blot the gel overnight and, after the transfer is completed, place the membrane in 2 × SSC/0.2 M Tris–HCl for 15 min with gentle shaking.
4. Place the membrane between filter paper and allow it to dry completely. No additional treatments are required before pre-hybridization.

B. Hybridization

1. Pre-hybridize the membrane in pre-hybridization mixture for at least 15 min at 55°C in a shaking water bath.
2. Add the radiolabelled DNA probe directly and incubate overnight at 55°C in a shaking water bath.
3. Remove the hybridization mixture and wash the filters three times in a large volume (100–200 ml) of 3 × SSC/0.5% SDS for 15–20 min.
4. Place the membrane on filter paper but do not allow it to dry completely otherwise the probe may bind irreversibly.
5. Keep the membrane moist by wrapping it in Saran® wrap before exposure to Kodak Royal X-omat film.
6. Expose the film with intensifying screens overnight at −70°C (see Chapter 3).

Protocol 5. Random primer labelling of GP-PCR probe (see also Chapter 2)

Materials

- 5 × oligo labelling mixture: for 500 μl, mix 100 μl of solution A, 250 μl of solution B, and 150 μl of solution C

solution A:	1.25 M Tris–HCl, 1.25 M MgCl$_2$, pH 8.0	1 ml
	β-mercaptoethanol	18 μl
	dTTP (100 mM)	5 μl
	dGTP (100 mM)	5 μl
	dATP (100 mM)	5 μl
solution B:	2 M Hepes, pH 6.6	
solution C:	hexanucleotides (90 OD Units per millilitre of TE buffer)	

Store at −20°C until use.

- [α-^{32}P]-dCTP at 3000 Ci/mmol (for example, NEN DuPont)
- TE buffer: 10 mM Tris–HCl, pH 7.5, 1 mM EDTA
- Low-melting-point agarose (BioRad)
- Sephadex G50
- Klenow DNA polymerase

Method

1. Perform GP-PCR (see *Protocol 2*) separately on 1 ng of cloned HPV 6, 11, 16, 18, 31, and 33 plasmid DNA.

2. Subject the PCR products to 1% low-melting-point agarose gel electrophoresis. (see Chapter 3 and *Protocol 1*, Chapter 5.)

3. Excise the GP-PCR fragments of 140–150 bp from the gel with as little agarose as possible, using razor blades.

4. Pool equal amounts of the excised fragments derived from the different cloned HPVs in a reaction tube.

5. Melt the agarose fragments by placing the tube for 10 min at 65°C and vortex vigorously.

6. Store at −20°C until use.

7. Re-melt the probe mixture at 65°C, add 4 μl of this probe mixture to 12 μl of distilled water and denature by boiling for 5 min.

8. Quickly add (do not allow the agarose to solidify) in the following order:
   - 5 × oligo labelling mixture   5 μl
   - [α-^{32}P]-dCTP at 3000 Ci/mmol   2 μl
   - Klenow DNA polymerase   1 μl

9. Incubate at 37°C for 30–60 min.

10. Add 100 μl of TE buffer, mix and place at 65°C for 10 min.

11. Separate the labelled probe from free nucleotides by Sephadex G50 filtration as follows:

12. Prepare a Sephadex G50 column in a siliconized Pasteur pipette (volume 3.5 ml).

13. Apply the probe sample to a column pre-equilibrated with TE buffer.

Protocol 5. *Continued*

14. Elute with 200 µl TE buffer five times, collecting each fraction in a separate reaction tube.

15. Determine the amount of radioactivity in each tube with a hand-held monitor.

16. Pool the last two fractions, which usually contain high amounts of labelled probe purified from free nucleotides.

17. Denature the probe by boiling for 5 min and add it to pre-hybridization mixture (see *Protocol 4*).

Protocol 6. 5′-end-labelling of TS-PCR cocktail probe (see also Chapter 2)

Materials

- 10 × kinase buffer: 500 mM Tris–HCl, pH 7.6, 100 mM MgCl$_2$, 50 mM DTT, 1 mM EDTA, 1 mM spermidine
- Type-specific oligonucleotide probes (10 pmol/µl) for each of HPV 6, 11, 16, 18, 31, and 33 (see *Table 2*)
- [γ-^{32}P]-dATP at 300 Ci/mmol (for example, NEN DuPont)
- T4 polynucleotide kinase

Method

1. Mix to a final volume of 20 µl:

• Each type-specific oligonucleotide probe	1 µl
• 10 × kinase buffer	2 µl
• Bidest	9 µl
• [γ-^{32}P]-dATP	3 µl
• T4 polynucleotide kinase	1 µl

2. Incubate for at least 30 min at 37°C.

3. Perform Sephadex G50 filtration as described in *Protocol 5* except that the 65°C heating step and the final denaturation step can be omitted.

4. Add the probe to the pre-hybridization mixture (see *Protocol 4*).

Weak GP-PCR hybridization signals may occur as a result of co-annealing of primers to cellular DNA and should not therefore be assigned as HPV X. Only GP-PCR positive products with strong hybridization signals (comparable to the signal obtained with 100 SiHa cells) and negative in the TS-PCR are considered as HPV X. This interpretation has been confirmed recently by sequence analysis of the amplified products (van den Brule *et al.*, manuscript submitted). To reduce false negativity, we periodically subject a randomly selected group of scrapes to PCR using β-globin specific primers (see *Table 2*) as an internal control for amplifiable target DNA (see also Chapters 4 and 6). In our experience, less than 0.1% of scrapes are β-globin PCR negative after gel electrophoresis, indicating sufficient smear quality for PCR purposes. The direct PCR is also fast: 120 samples can be processed for PCR within 2 h. In this way, one technician can routinely screen 500 samples of a patient population per week for HPV using the whole procedure. Rapidity is not restricted by sample pre-treatment and can be enhanced by shortening of hybridization and autoradiography. Application of non-radioactive detection of PCR products and automation will further increase the use of PCR in routine HPV screening of cervical scrapes.

Although two important problems for the application of PCR to routine mass screening, namely the omission of DNA purification and the detection of unsequenced HPV genotypes, have been solved, other PCR problems such as contamination (see Chapters 4 and 6) must be kept in mind.

3.3 Some critical remarks concerning HPV screening

Different generalized HPV PCR methods have recently been described, using general or consensus primers located in the E1 (18, 27) and L1 (16, 23) ORFs. Since the largest nucleotide homology amongst the sequenced HPVs does not exceed 12 consecutive nucleotides, all groups have adapted either the primers or the method to ensure amplification of a broad spectrum of HPVs. Adaptations include a certain degree of mismatch acceptance between primers and target DNA accomplished by reducing the stringency of primer annealing (ref. 16, and this chapter), the use of degenerate primers (a mixture of oligonucleotides showing nucleotide differences at several positions) to render them sufficiently complementary to HPV 6, 11, 16, 18, and 33 (see Chapter 6), and the use of primers which in addition contain inosine residues at certain ambiguous base positions (27). The HPV consensus PCR with the MY 09/11 primer combination described by Manos *et al.* (1989) (ref. 23, and see Chapter 6) has also been applied successfully to clinical samples (28, 29). However, comparison of HPV 16 amplification of the GP 5/6 system with the originally described MY 09/11 protocol, which uses 30 cycles of amplification and a generic oligonucleotide probe for hybridization (23), revealed that the latter protocol was at least 10 times less sensitive. This is likely to be due to differences in the number of cycles performed (40 versus 30) and may also reflect the different probes used (mixture of full length amplification products versus generic oligonucleotide probe). Recently, the MY 09/11 system has been adapted by an increased number of cycles and the use of a cocktail probe of PCR products, internally amplified from several cloned and unknown HPV's (29). However, we favour the GP 5/6 PCR because of the shorter length of generated amplification fragments (150 versus 450 bp), which may be advantageous under less efficient amplification conditions such as crude cell suspensions. Furthermore, it is recommended that different laboratories exchange cervical scrapes used for HPV detection by a standardized PCR method for a consensus statement about HPV prevalence in cytological smears.

The main drawback of the PCR is its sensitivity, as clinical or laboratory-generated contamination by environmental DNA, cloned plasmids and PCR products may produce false-positive results. In order to avoid false-positives from plasmid HPVs (pHPVs), the use of anticontamination or cloning-site flanking primers was introduced (see *Figure 5*). Application of these primers makes it possible to detect HPVs specifically in a clinical sample even in the presence of pHPVs (17). In our experience, however, amplification products are the most serious source of contamination at present and have forced us to carry out the different PCR steps, such as sample preparation, electrophoresis, and PCR solution preparation in three spatially separated rooms (see also Chapters 4 and 6). This requires strong laboratory discipline. During the whole procedure, the use of disposable tubes and special pipette tips is also necessary (see *Protocol 1*). By direct performance of PCR on crude samples, the number of pretreatment steps is greatly reduced. With these precautions, the direct HPV-GP/TS-PCR is reliable to perform and has the advantage that contamination can be minimized.

4. Clinical aspects

4.1 HPV prevalence in relation to cytology

Epidemiological studies have revealed great variation in HPV prevalence

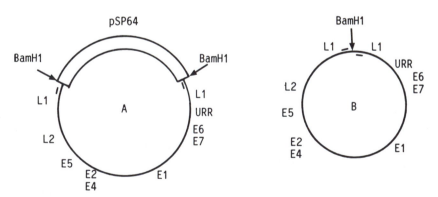

Figure 5. Principle of anticontamination primers. Bars indicate primers flanking the unique *Bam*H1 cloning site, and (A) direct the amplification of the 3 kb pSP64 fragment, or (B) direct the amplification of a 200 bp fragment of the episomal HPV-DNA. The PCR conditions used in the HPV TS-PCR are optimal for the amplification of up to a 500 bp fragment. E = early genes; L = late genes.

rates due to several factors, such as the different detection techniques used and the composition of the study groups tested (30). The HPV GP/TS-PCR method is the appropriate tool for investigating the prevalence of HPV in cytological smears. Van den Brule *et al.* (1991) (20) applied the method to two different groups of women:

(a) a symptom-free population consisting of women aged 30–35 years living in a region with a low incidence of cervical cancer and involved in triennial screening for cervical cancer;

(b) a gynaecological out-patient population containing women visiting the clinic for a wide spectrum of gynaecological complaints both with and without a history of cervical pathology (aged 16 to 60 years).

Routine cytological screening of smears and PCR detection were performed independently. Significant differences in overall HPV prevalence rates in the cytologically normal scrapes of asymptomatic women (A–I; 3.5%), patients with no history of cervical pathology (B–II; 9%), and patients with a history of cervical pathology (C–II; 21.5%) were found as shown in *Figure 6*. The frequencies of HPV types 16 and 18 in these populations were 0.9% (A), 2.5% (B), and 12% (C) respectively. In the out-patient group, 78% of smears containing HPV 16 and 18 were associated with a history of cervical intraepithelial neoplasia grades 1 to 3. This may imply that women with cytologically normal cervical smears containing oncogenic HPV types should be followed up clinically. Such a strategy is now being carried out in our hospital. In smears with mild and severe dysplasia and smears suspected of carcinoma *in situ* from both populations, the overall HPV prevalence was 70%, 84%, and 100% respectively. In all squamous cell carcinoma of the cervix (*n* = 50), HPV was detected. Frequencies of HPV 16 and 18 increased from 41% in mild dysplasia to 94% in cervical carcinomas.

4.2 HPV detection and cytological screening in cervical cancer: future developments

HPV has been found in all invasive carcinomas and carcinomas *in situ* tested to date (see *Figure 6*). Thus, the presence of HPV can be used as a marker for neoplastic cells. The presence of HPV in a considerable percentage (70%) of cervical smears with dysplasia and the low percentage of HPV (3.5%) in cytologically normal smears suggests that the presence of HPV in these smears detects a population of women who are at increased risk of developing cervical cancer. Also, the detection of potentially oncogenic HPV types using the GP/TS-PCR assay might predict those cervical lesions that progress to invasive carcinoma. At the moment, this hypothesis is being tested in a large clinical follow-up study in our hospital. The implication of this assumption is that screening for HPV in cervical smears could replace cytological screening for abnormal cells in large-scale screening programmes for cervical cancer

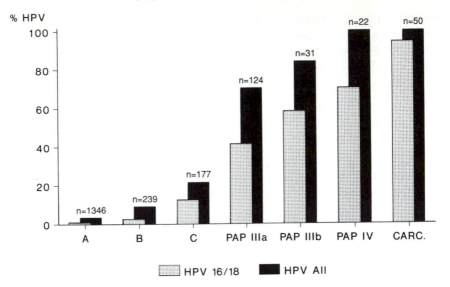

Figure 6. HPV positivity in relation to cytological analysis of cervical smears and to tissue specimens of cervical carcinomas (Carc). Overall HPV positivity (HPV All), including still unsequenced HPV genotypes, was determined by general primer-mediated PCR, while HPV types 16 and 18 were detected by type-specific PCR. Cytologically normal scrapes (Pap I, II) of different populations A (asymptomatic group), B and C (gynaecological out-patient population), as described in the text, and cytologically abnormal scrapes (Pap IIIa, IIIb, and IV) were studied for the presence of HPV. HPV was detected in crude cell suspension of cervical scrapes and in DNA purified from tissue samples of squamous-cell carcinomas (according to Walboomers *et al.* 1988, ref. 37).

in women older than 35 years. Only women with HPV positive smears would then be followed up cytologically to detect the patients with cytomorphologic-ally abnormal cells. A considerable number of HPV-negative women with dysplastic cervical cells would then not be treated by gynaecologists.

Preliminary results from our follow-up study suggest that progression is always associated with the presence of HPV, predominantly types 16 and 18. Since HPV screening, as described in this chapter, seems much more repro-ducible and less time-consuming than the screening of cytological smears, the results of ongoing clinical follow-up studies have important implications for the way in which cervical screening is carried out in the near future.

Acknowledgements

The authors wish to thank Miss Yvonne Duiker and Mrs Carla van Rijn for the excellent preparation of the manuscript and Dr P. van der Valk for critical reading. The cloned HPV types used in this study were kindly provided by Drs E. M. de Villiers, H. zur Hausen, L. Gissmann (Heidelberg, Germany),

A. Lorincz (Gaithersburg, Maryland, USA), K. Shah (Baltimore), G. Nuovo (New York) and G. Orth (Paris, France). This work was in part supported by the Dutch Cancer Society Koningin Wilhelmina Fonds: Grants IKA-VU-89-16; IKA-VU-91-17 and the Prevention Fund, The Netherlands, Grant 28-1502.2.

References

1. Richart, R. M. and Barron, B. A. (1981). *Cancer,* **47,** 1176
2. Campion, H. J., Cuzick, J., McCance, D. J., and Singer, A. (1986). *Lancet,* **2,** 237.
3. Koss, L. G. (1989). *J. Am. Med. Assoc.,* **261,** 737.
4. Gissmann, L. and Schneider, A. (1986). *Herpes and papilloma viruses* (ed. G. De Palo, F. Rilke, and H. zur Hausen), pp. 15–25. New York Serzona Symposia Publications. Raven Press, New York.
5. zur Hausen, H. (1989). *Cancer Res.,* **49,** 4677.
6. Gissmann, L., Wolnik, L., Ikenberg, H., Koldowsky, U., Schnurch, H. G., and zur Hausen, H. (1983). *Proc. Natl. Acad. Sci. USA,* **80,** 560.
7. Lorincz, A. T., Lancaster, W. D., Kurman, R. J., Bennett Jenson, A., and Temple, G. F. (1986). In *Viral etiology of cervical cancer* (ed. R. Peto and H. zur Hausen), p. 225. Cold Spring Harbor Laboratory, Cold Spring Harbor, NY.
8. Lorincz, A. T., Lancaster, W. D., and Temple, G. F. (1986). *J. Virol.,* **58,** 225.
9. McCance, D. J., Campion, M. J., Clarkson, P. K., Chesters, P. M., Jenkins, D., and Singer, A. (1985). *Br. J. Obstet. Gynaecol.,* **92,** 1101.
10. Boshart, M., Gissmann, L., Ikenberg, H., Kleinheinz, A., Scheurlen, W., and zur Hausen, H. (1984). *EMBO J.,* **3,** 1151.
11. Durst, M., Gissmann, L., Ikenberg, H., and zur Hausen, H. (1983). *Proc. Natl. Acad. Sci. USA,* **80,** 3812.
12. Beaudenon, S., Kremsdorf, D., Croissant, O., Jablonska, S., Wain-Hobson, S., and Orth, G. (1986). *Nature,* **321,** 246.
13. DeVilliers, E. M. (1989). *J. Virol.,* **63,** 4898.
14. Saiki, R. K., Scharf, S., Faloona, F., Mullis, K. B., Horn, G. T., Erlich, H. A., and Arnheim, N. (1985). *Science,* **230,** 1350.
15. Saiki, R. K., Gelfand, D. H., Stoffel, S., Scharf, S. J., Higuchi, R., Horn, G. T., Mullis, K. B., and Ehrlich, H. A. (1988). *Science,* **239,** 487.
16. Snijders, P. J. F., van den Brule, A. J. C., Schrijnemakers, H. F. J., Snow, G., Meijer, C. J. L. M., and Walboomers, J. M. M. (1990). *J. Gen. Virol.,* **71,** 173.
17. van den Brule, A. J. C., Claas, H. C. J., du Maine, M., Melchers, W. J. G., Helmerhorst, T., Quint, W. G. V., Lindeman, J., Meijer, C. J. L. M., and Walboomers, J. M. M. (1991). *J. Med. Virol.,* **29,** 20.
18. van den Brule, A. J. C., Snijders, P. J. F., Gordijn, R. L. J., Bleker, O. P., Meijer, C. J. L. M., and Walboomers, J. M. M. (1990). *Int. J. Cancer,* **45,** 644.
19. van den Brule, A. J. C., Meijer, C. J. L. M., Bakels, V., Kenemans, P., and Walboomers, J. M. M. (1990). *J. Clin. Microbiol.,* **28,** 2739.
20. van den Brule, A. J. C., Walboomers, J. M. M., du Maine, M., Kenemans, P., and Meijer, C. J. L. M. (1991). *Int. J. Cancer,* **48,** 404.
21. Li, H., Gyllensten, U. B., Cui, X., Saiki, R. K., Erlich, H. A., and Arnheim, N. (1988). *Nature,* **335,** 414.

22. Kogan, S. C., Doherty, M., and Gitschier, J. (1987). *New Engl. J. Med.*, **317,** 985.
23. Manos, M. M., Ting, Y., Wright, D. K., Lewis, A. J., Broker, T. R., and Wolinsky, S. M. (1989). *Cancer Cells,* **7,** 209.
24. Melchers, W., van den Brule, A., Walboomers, J., de Bruin, M., Herbrink, P., Meijer, C., Lindeman, J., and Quint, W. (1989). *J. Med. Virol.,* **27,** 329.
25. Queen, C. and Korn, L. J. (1984). *Nucleic Acids Res.,* **12,** 581.
26. Giri, I. and Danos, O. (1986). *Trends in Genet.,* **2,** 227.
27. Gregoire, L., Arella, M., Campione-Piccardo, J., and Lancaster, W. D. (1989). *J. Clin. Microbiol.,* **27,** 2660.
28. Resnick, R. M., Cornelissen, M. T. E., Wright, D. K., Eichinger, G. H., Fox, H. S., ter Schegget, J., and Danos, M. M. (1990). *J. Natl. Cancer Inst.,* **82,** 1477.
29. Bauer, H. M., Ting, Y. I., Greer, C. E., Chambers, J. C., Tashiro, C. J., Chimera, J., Reingold, A., and Manos, M. (1991). *JAMA,* **265,** 472.
30. Munoz, N., Bosch, X., and Kaldor, J. M. (1988). *Br. J. Cancer,* **57,** 1.
31. Schwarz, E., Durst, M., Demankowski, C., Lattermann, O., Zech, R., Wolf-sperger, E., Suhai, S., and zur Hausen, H. (1983). *EMBO J.,* **2,** 2341.
32. Dartmann, K., Schwarz, E., Gissmann, L., and zur Hausen, H. (1986). *Virol.* **151,** 124.
33. Seedorf, K., Krammer, G., Durst, M., Suhai, S., and Rowekamp, W. G. (1985). *Virol.,* **145,** 181.
34. Cole, S. T. and Danos, O. (1987). *J. Mol. Biol.,* **193,** 599.
35. Goldsborough, M. D., Di-Silvestre, D., Temple, G. F., and Lorincz, A. T. (1989). *Virol.,* **171,** 306.
36. Cole, S. T. and Streeck, R. (1986). *J. Virol.,* **58,** 991.
37. Walboomers, J. M. M., Melchers, W. J. G., Mullink, H., Meijer, C. J. L. M., Struyk, A., Quint, W. G. J., van der Noordaa, J., and ter Schegget, J. (1988). *Am. J. Pathol.,* **131,** 587.

8

Molecular detection and identification of pathogenic organisms

BRUCE J. McCREEDY and JOSEPH A. CHIMERA

1. Introduction

During the last decade an increasing body of literature has accumulated demonstrating the utility of DNA probes as diagnostic tools for the clinical laboratory (reviewed in refs. 1 and 2). Because the genome of each micro-organism contains sequences that are unique to the species, accurate identification can be achieved for any pathogen using properly designed DNA probes. For the clinical laboratory, DNA probe technology offers the opportunity to identify rapidly organisms that are difficult to identify using culture techniques. Direct detection of pathogens in primary clinical specimens using DNA probes has significantly reduced the time and expense associated with the isolation and identification of many infectious agents.

DNA probe technology is based upon the highly specific nature of binding between complementary base-pairs of two molecules of DNA. The double-stranded DNA molecule of the pathogen can be denatured by heat or certain chemicals yielding single-stranded DNA. A nucleic acid probe, consisting of a segment of DNA (or RNA), modified by the incorporation of a reporter molecule, can then seek out and bind to complementary bases present within a target nucleic acid sequence. The target sequences are provided by the nucleic acid of an infectious agent present in the specimen. Reporter molecules for nucleic acid probes may be radioactive molecules, enzymes, or various types of ligands such as biotin molecules (see Chapter 2). Hybridization is the process of association of two strands of nucleic acid as a result of binding between complementary base-pairs on either strand. Binding of a DNA probe with a target nucleic acid sequence occurs during an *in vitro* hybridization reaction. Hybridization reactions can be performed using several formats including nucleic acid immobilized of the target DNA on a solid support, in solution, and *in situ* (see Chapter 3 and Volume I). Depending upon the nature of the specimen and suspected pathogen, and the type of probe used, different hybridization formats are used to achieve the most sensitive and reliable detection of target DNA sequences.

2. Design and selection of probes

Nucleic acid probes for detecting pathogenic organisms may be derived from several different sources but all have the common property of being specific for only the organism(s) which the probe is designed to detect. This is accomplished by designing probes that are complementary to unique sequences within the genome of an organism, and can be genus- or species-specific. Probes may consist of relatively long stretches (10^2–10^4 base-pairs) of genomic DNA obtained by molecular cloning of portions of the pathogen's genome. Long probes of single-stranded nucleic acid can also be obtained by *in vitro* transcription reactions which generate highly specific RNA probes (see Chapter 2). Synthetic oligonucleotide probes, which are chemically synthesized single-stranded DNA molecules, usually between 12 and 100 bases in length, are now commonly used in clinical laboratories, and many are commercially available for the detection of specific pathogens. Hybridization reactions using short oligonucleotide probes are highly specific and may proceed very rapidly in comparison with longer probes. However, decreased sensitivity has been reported for some oligonucleotide probes in comparison with longer probes due to the fewer number of reporter molecules that can be incorporated. Some recently introduced amplification systems have greatly increased the level of sensitivity obtained using oligonucleotide probes. Selection of probes should therefore be based upon several factors including the availability of commercially prepared probes or the capability to synthesize an oligonucleotide probe. The level of sensitivity and specificity required to achieve the detection goals must also be considered when selecting probes, reporter molecules, and hybridization conditions. Different methods for performing hybridization reactions, selection and labelling of probes, and selection of reporter molecules are described in detail in an earlier volume of this series (3).

3. Detection of enteric pathogens by colony hybridization assay

Diarrhoeal diseases are the cause of millions of deaths each year in developing countries. Several different bacterial and viral pathogens can cause these infections, necessitating prompt identification of the specific pathogen responsible for an outbreak of disease. Some enteric pathogens, such as enterotoxigenic *Escherichia coli* (ETEC), cause disease via the production of toxins that stimulate hypersecretion by intestinal epithelial cells. Other intestinal pathogens, such as several species of *Shigella*, are responsible for bacillary dysentery through direct invasive destruction of colonic epithelial cells. Normally, identification of these organisms is achieved by *in vitro* cultivation of stool specimens followed by biochemical tests and agglutination assays. ETEC must be further differentiated from normal commensal *E. coli* by detection of toxin production using specific antisera and tissue culture toxicity assays. Detection of these pathogens using DNA probes, following growth in selective media or directly from stool specimens, greatly reduces the time and expense required to identify the specific organism responsible for the disease.

3.1 Preparation of specimens for hybridization analysis

Protocol 1 describes a method for transfer of colonies from the surface of selective growth media to nylon or nitrocellulose membranes, followed by lysis of the colonies and binding of the liberated DNA to the membrane. The method is an adaptation of the procedure originally described by Grunstein and Hogness (4).

Protocol 1. Colony transfer to solid supports for hybridization following growth in selective media

Materials
- 10% sodium dodecyl sulphate (SDS)
- Denaturing solution: 0.5 M NaOH, 1.5 M NaCl
- Neutralizing buffer: 0.5 M Tris–HCl, pH 8.0, 1.5 M NaCl.

Method
1. Inoculate a stool specimen or rectal swab on to MacConkey and Hektoen agar plates.[a] Invert the plates and incubate them at 37°C for 12–24 h, periodically observing for colony growth.

2. Place a nylon or nitrocellulose filter (82 mm circle) on to the surface of a nutrient agar plate.[b] Wear gloves when handling filters.

3. Using sterile toothpicks or disposable inoculating loops, transfer colonies from the selective growth plates to the filter. Make small streaks when inoculating the filter, keeping transferred colonies separated.

4. In a separate area of the filter, transfer a colony of positive control cells, and a colony of negative control cells (that is, non-toxigenic *E. coli*) from overnight cultures. Be sure to mark the filter clearly with the location of the positive and negative reference cells. This can be done by piercing holes in the filter with a 20-gauge needle or by marking the membrane with indelible ink. The positive control cells contain the nucleic acid sequence for which the probe is specific (that is, a portion of the toxin-encoding gene) while the negative control cells lack this sequence.

175

Protocol 1. *Continued*

5. Invert the plates and incubate them at 37 °C for 12 h.

6. Remove the filter from the agar plate using blunt-ended forceps and place it 'colony side up' on to several sheets (3–4) of Whatman 3MM filter paper which have been pre-soaked in 10% SDS for 5 min.

7. Blot excess fluid from the filter and transfer to a second stack of 3MM paper saturated with denaturing solution for 5 min.

8. Transfer the filter to 3MM paper pre-soaked with neutralizing buffer for 5 min.

9. Transfer the filter to a dry piece of 3MM paper and let the filter dry at room temperature for 37 °C for 30–60 min.

10. Place the filter between two pieces of 3MM paper and place between two glass plates. Bake for 2 h at 80 °C in a vacuum drying oven. The denatured nucleic acid can also be permanently fixed to the membrane by ultraviolet (UV) crosslinking in a matter of minutes using commercially available UV emitting devices (for example, Stratalinker, Stratagene, Inc.). In our experience this method has yielded good results with charged nylon membranes, but was less efficient when using nitrocellulose membranes. The filters can then be used immediately in a hybridization reaction, or stored wrapped in aluminium foil under vacuum at room temperature.

[a] Direct detection from stool specimens is also possible although less sensitive. Using this method, a primary specimen is streaked or spread on to a nylon or nitrocellulose membrane and placed on the surface of a nutrient agar plate. Following incubation for 24 h at 37 °C, the colonies are prepared for hybridization analysis as described in steps 6–10.

[b] Details of selective and non-selective agar plate preparation are given in a further volume of this series (5).

3.2 Hybridization of nucleic acid bound to nitrocellulose or nylon filters with enzyme-labelled DNA probes

The conditions under which hybridization will proceed optimally varies according to the type of probe used (that is, long or short, double- or single-stranded, etc.). The conditions required for detection of hybridized probe also vary according to the type of reporter molecule attached to the probe. Many reviews are available which discuss the different parameters affecting hybridization and detection reactions according to the type of probe and label used. The following protocol describes the use of short oligonucleotide probes containing covalently attached alkaline phosphates (AP) moieties. Alkaline phosphatase labelled probes can be prepared as described (6) or purchased from commercial suppliers (for instance, Cambridge Research Biochemicals, UK).

Protocol 2. Hybridization of nucleic acid immobilized on filters to alkaline phosphatase (AP) labelled DNA probes and detection by dye precipitation

Materials

- Pre-hybridization solution: 5 × SSC, 1% (v/v) SDS, 0.5% (w/v) bovine serum albumin, 0.5% (w/v) polyvinylpyrrolidine
- 20 × SSC: 3 M NaCl, 0.3 M sodium citrate, pH 7.0
- Wash buffer: 0.1 M Tris–HCl, pH 8.5
- Nitroblue tetrazolium (NBT): 75 mg/ml stock solution in 70% dimethylformamide (DMF)
- 5-bromo-4-chloro-3-indolyl phosphate (BCIP): 50 mg/ml stock solution in DMF
- Substrate buffer: 0.1 M Tris–HCl, 0.1 M NaCl, 50 mM $MgCl_2$, pH 8.5

Method

1. Place filters containing fixed target nucleic acid (see *Protocol 1*) into heat-sealable plastic pouches and pre-hybridize in pre-hybridization solution at 50°C for 15 min. Add sufficient pre-hybridization solution to cover the membrane and allow it to float freely (for example, 100–200 μl/cm^2).

2. Cut off a corner of the pouch and remove the pre-hybridization solution. Add fresh pre-hybridization solution containing 50 ng of AP-labelled probe. Use only as much hybridization solution as is necessary to recover the membrane. Re-seal the pouch and allow hybridization to proceed for 15–30 min at 50°C.

3. Wash filters in two changes of 10 ml of 1 × SSC, 1% SDS at 50°C for 5 min each.

4. Wash the filters once for 5 min in 1 × SSC without detergent. SDS and other detergents can interfere with the precipitation of dye during the colour development reaction.

5. Develop the filters in a solution of 0.33 mg/ml NBT, 0.17 mg/ml BCIP in substrate buffer at room temperature until fully developed. Colour development is usually complete within an hour.

Detection of target nucleic acid by the probe is indicated by precipitation of dye, resulting in a purple colour at the site of the colony on the filter. Discrimination of true positive samples from negative samples is accomplished by comparison with the reactions observed on the filter for the positive and negative control colonies. Some background colour is normal: background reactivity can be distinguished from weakly reactive samples by comparison of the sample with the colour observed for the negative control colony.

4. Detection of pathogens in clinical specimens using polymerase chain reaction (PCR) amplification

The ability to diagnose a variety of infections of bacterial and viral origin rapidly by direct examination of a clinical specimen has suffered due to a lack of sensitivity of the methods used for detection. The advent of molecular biological techniques employing gene probes for identification of pathogenic organisms has provided several advantages over the classical techniques of *in vitro* cultivation followed by biochemical or serological testing. Although the use of DNA probes can circumvent the need to perform biochemical tests and immunological assays, probe technology has often been shown to be too insensitive for use in the direct identification of organisms in complex clinical specimens such as stool and urine. Therefore, the need to culture pathogens prior to identification with nucleic acid probes has limited the use of probe technology in diagnostic medicine. The introduction of PCR amplification, coupled with gene probe detection, has provided the sensitivity and specificity required for the direct detection of target nucleic acid sequences in clinical specimens. The polymerase chain reaction is a powerful technique that can produce millions of copies of a selected DNA target in only a few hours. By using PCR, it is possible to amplify and detect selectively as few as one target DNA molecule in a complex sample of genomic DNA (7). Several groups have demonstrated the utility of PCR coupled with DNA probes for direct detection of pathogens in clinical specimens (see refs. 8, 9, and 10; an example of this application is given in *Protocol 4*). For a review of the principles and methodologies involving the use of PCR for a variety of clinical and research applications, see Innis *et al.* (11) and also Chapters 4, 6, 7, 9, and 10 of this volume.

Protocol 4. Polymerase chain reaction amplification for the detection of bacterial pathogens in clinical specimens

Materials
- Transport medium: 0.02 M phosphate buffer, pH 7.2 containing 0.2 M sucrose, 10% heat-inactivated fetal calf serum, 25 µg/ml gentamicin and 2.5 µg/ml amphotericin B
- Digestion buffer: 50 mM Tris–HCl, pH 8.5, 1 mM EDTA, 1% (v/v) SDS, 200 µg/ml proteinase K
- Light mineral oil
- PCR reaction buffer: 10 mM Tris–HCl, pH 8.3 at 25°C, 50 mM KCl, 2.5 mM $MgCl_2$, 0.01% gelatin and 200 µM each of dATP, dCTP, dGTP, and dTTP

Method

1. Clinical specimens such as cervical and urethral swabs should be placed directly into transport medium. Urine specimens less than 24 h old can be used directly. Stool specimens should be re-suspended in a physiological buffer (for example, phosphate-buffered saline: 10 mM phosphate, 150 mM NaCl, pH 7.4) to a concentration of 200–500 mg/ml.

2. Remove insoluble particles by low speed centrifugation (200 g for 5 min). Remove 0.5 ml of the clarified solution and pellet the bacteria by centrifugation in an Eppendorf (or equivalent) centrifuge at 12 000 g for 1 min. Re-suspend the pellets in 100 μl of digestion buffer.[a]

3. Mix the samples by vortexing to disrupt the pellets and incubate at 55°C for 1 h.

4. Inactivate the proteinase K by incubating the samples at 95°C for 10 min. Spin the samples briefly (2 sec) in a microcentrifuge to bring the condensation to the bottom of the tubes.

5. Label 0.5 ml GeneAmp (Perkin–Elmer Cetus) reaction tubes with each sample identification and add 75 μl of light mineral oil. Add 100 μl of 1 × PCR reaction buffer containing 0.2 μM of each primer, and 2.5 U of *Taq* DNA polymerase per 100 μl (Amplitaq, Perkin–Elmer Cetus).[b]

6. Add 1–10 μl of sample from step 4 to the labelled tubes containing mineral oil and 1 × PCR buffer from step 5. Mix and spin in a microcentrifuge for 10 sec to bring the aqueous phase under the oil.

7. Place the samples into an automated DNA thermal cycler (from, for example, Perkin–Elmer Cetus, Norwich, Connecticut, USA) and perform 30–35 cycles. Each cycle consists of denaturation at 95°C for 1 min, annealing at 55°C for 30 sec and extension at 72°C for 1 min. During the last cycle, an additional extension time of 5 min should be included. Positive and negative control DNAs should be included with each PCR reaction to ensure that no inhibitors or contaminating DNAs have been introduced during the sample preparation or by the PCR reaction components.

8. If samples cannot be analysed for amplified target DNA sequences immediately, programme the thermal cycler to hold the samples at 15°C until analysis can be performed, or store the amplified products at 4°C.

---

[a] Different organisms may require different lysis conditions. Additional components such as lysozyme may be necessary in order to liberate the nucleic acid from some organisms (see Chapter 2). Phenol–chloroform extraction and ethanol precipitation of nucleic acids may be required for some samples if no PCR products are obtained directly from the lysed sample (see Chapter 1).
[b] The PCR reaction buffer components must be optimized for each individual system with respect to the concentrations of magnesium ions, primers, *Taq* DNA polymerase, and dNTPs (see Chapter 4). However, for most reactions, the conditions listed above will yield PCR amplification products in positive clinical specimens.

Analysis of PCR amplification products can be accomplished using several formats. PCR products can be immobilized on to solid support membranes and hybridized to labelled nucleic acid probes. A solution hybridization reaction using radiolabelled probes followed by gel electrophoresis and autoradiography is also a commonly used method of analysis. Due to the amplification power of PCR, amplified target can often be visualized directly in ethidium-bromide-stained agarose gels. However, additional sensitivity and specificity are obtained using DNA probes for analysis and detection of PCR products. A rapid method for analysing amplification products using a dot blot format and probing with ^{32}P-labelled probes is described in *Protocol 5*. This protocol describes the use of a dot-blot manifold, but the application of samples to filters by hand is equally effective (see Chapter 3).

Protocol 5. Analysis of PCR amplification products by preparation of dot blots and hybridization with ^{32}P-labelled probes

Materials
- Denaturing solution: 0.4 M NaOH, 25 mM EDTA. This should be made fresh.
- 20 × SSPE: 3.6 M NaCl, 0.2 M sodium phosphate, 0.11 M NaOH, 0.02 M EDTA, pH 7.4.
- Hybridization buffer: 2 × SSPE, 0.5% (v/v) SDS, 5 × Denhardt's solution.
- 50 × Denhardt's solution: 1% (w/v) Ficoll, 1% (w/v) polyvinylpyrrolidine, 1% (w/v) bovine serum albumin.
- Blot wash solution: 2 × SSPE, 0.1% (v/v) SDS.

Method
1. Pre-wet a 0.45 μm nylon membrane (BioDyne B, New York, USA) by immersing it in distilled water for 20–30 min. Orientate the membrane by cutting a notch in one corner and place it into a dot-blot apparatus (for example, Bio-dot, Bio-Rad, California, USA). Secure the membrane in the apparatus, making sure to remove any trapped air-bubbles.
2. Label a sufficient number of microtubes for the amplified samples.[a] For each sample to be analysed (including controls) add 95 μl of fresh denaturation solution to the corresponding tube.
3. Add 5 μl of PCR product to the appropriately labelled tubes and mix the contents. Incubate for 10 min at room temperature to allow the amplification products to denature.
4. Add 100 μl of denaturation solution to all wells of the dot-blot apparatus and gently filter through by applying the vacuum. Do not leave the wells completely dry.

5. Load 100 μl of each denatured sample into the appropriate well of the dot-blot apparatus, being careful to avoid trapping air-bubbles in the wells.

6. When all the samples have been added, apply the vacuum and draw the solution through. Remove any bubbles with a clean pipette tip to allow all of the solution to be drawn through the membrane. Return the apparatus to atmospheric pressure.

7. Add 200 μl of 20 × SSPE to each well. Apply vacuum and filter the solution through.

8. Seal off the vacuum line to trap the vacuum in the dot blot apparatus, then turn off the vacuum source. Remove the top of the apparatus and label the membrane with the sample information using a coloured pencil. Using blunt-ended forceps, remove the membrane and place it on a piece of Whatman 3MM filter paper soaked in 2 × SSPE.

9. Fix the filtered products on to the membrane by placing it 'sample-side-up' into a UV crosslinking apparatus (for example, Stratalinker, Strata-gene, Inc.) and setting the instrument for 120 000 μjoule.[b]

10. Rinse the membrane in 2 × SSPE and store in a heat-sealable plastic bag containing 20 ml of 2 × SSPE unless the hybridization reaction is to be performed immediately.

11. Pre-warm hybridization buffer and blot wash solution in a 65 °C water bath.

12. Place the membrane in a heat-sealable bag and add 20 ml of pre-warmed hybridization buffer. Two membranes can be placed together 'DNA-sides-out' and placed into the same bag.

13. Pre-hybridize the membranes by incubating them in a 65 °C water bath for at least 30 min. This serves to block sites on the membrane that might bind the labelled probe non-specifically and yield high background levels.

14. Cut off a corner of the bag and remove the pre-hybridization solution. Replace it with fresh pre-warmed hybridization buffer containing 1 pmol/ml (5.0×10^5–1.0×10^6 c.p.m./pmol) of ^{32}P-labelled oligonucleotide probe (see Chapter 2). Remove as many bubbles as possible from the bag then re-seal it and hybridize at 60 °C for 2 h. The probe is designed to recognize a sequence present in the amplified target DNA that is internal to the regions defined by the PCR primers used in the amplification reaction. Methods for producing ^{32}P end-labelled probes using T4 poly-nucleotide kinase are described in Chapter 2 (see refs 3 and 12).

15. Remove the membrane from the bag and dispose of the radioactive hybridization solution appropriately. Rinse the membrane briefly with blot wash solution at room temperature then wash at 60 °C for 15 min with further pre-warmed blot wash solution.

Protocol 5. *Continued*

16. Dry the membrane by placing on Whatman 3MM paper at room temperature or 37°C for 1 h.

17. Wrap the membrane in cling wrap and expose to X-ray film at −70°C with an intensifying screen (see Chapter 3). Develop according to the film requirements. Exposure times will vary according to the specific activity of the probe and the degree of hybridization that has occurred between probe and target.

---

a Alternatively, a sterile 96-well microtitre plate can be labelled and used to dilute samples in denaturation solution.
b Baking the membrane under vacuum at 80°C for 1–2 h is also acceptable if a UV crosslinker is not available or nitrocellulose membranes are used.

References

1. Kingsberry, D. T. and Falkow, S. (1987). In *Rapid detection and identification of infectious agents*. Academic Press, Orlando, Florida.
2. Tenover, F. C. (1988). *Clin. Microbiol. Rev.*, **1**, 82.
3. Hames, B. D. and Higgins, S. J. (ed.) (1985). *Nucleic acid hybridization: a practical approach*. IRL Press, Oxford.
4. Grunstein, M. and Hogness, D. (1975). *Proc. Natl. Acad. Sci. USA*, **72**, 3961.
5. Hawhey, P. M. and Lewis, D. A. (ed.) (1989). *Medical bacteriology: a practical approach*. IRL Press, Oxford.
6. Jablonski, E., Moomaw, E. W., Tullis, R. H., and Ruth, J. L. (1986). *Nucleic Acids Res.*, **14**, 6115.
7. Saiki, R. F., Gelfand, D. H., Stoffel, S., Scharf, S. J., Higuchui, R., Horn, G. T., Mullis, K. B., and Erlich, H. A. (1988). *Science*, **239**, 487.
8. Frankel, G., Giron, J. A., Valmassoi, J., and Schoolnik, G. K. (1989). *Mol. Micro.*, **3**, 1729.
9. Claas, H. C. J., Wagenvoort, J. H. T., Niesters, H. G. M., Tio, T. T., Van Rijsoort-Vos, J. H., and Quint, W. G. V. (1991). *J. Clin. Micro.*, **29**, 42.
10. Jensen, J. S., Uldum, S. A., Sondergard-Andersen, J., Vuust, J., and Lind, K. (1991). *J. Clin. Micro.*, **29**, 46.
11. Innis, M. A., Gelfand, D. H., Sninsky, J. J., and White, T. J. (ed.) (1990). *PCR protocols, a guide to methods and applications*. Academic Press, San Diego, California.
12. Maniatis, T., Fritsch, E. F., and Sambrook, J. (ed.) (1982). *Molecular cloning, a laboratory manual*. Cold Spring Harbor Press, Cold Spring Harbor, NY.

9

Human identification by DNA analysis

The amplification fragment length polymorphism analysis technique

MARCIA EISENBERG and JOSEPH A. CHIMERA

1. Introduction

Over the past several years, techniques that rely on the detection of genetic differences among individuals have become an increasingly common method to provide objective evidence in the fields of paternity testing and forensic science. Genetic variability can be detected by analysing the nucleic acid structure of highly polymorphic loci, specifically, stretches of DNA that contain variable numbers of short repetitive sequences referred to as VNTRs (variable number of tandem repeats). Analysing the allelic pattern of several polymorphic VNTR loci in an individual can yield probabilities for investigating biological relationships or for matching forensic material found at a crime scene to a suspected felon.

Traditionally, the restriction fragment length polymorphism (RFLP) method has been used to detect and analyse VNTR sequences. This approach involves isolating purified DNA, fragmenting the DNA with a restriction endonuclease, performing analytical gel electrophoresis on the restricted DNA, transferring the DNA from the gel to a nylon membrane, performing DNA blot hybridization with a radiolabelled probe containing the VNTR sequence of interest, and is concluded by autoradiography. Generally, RFLP analysis requires five to ten days to perform when it is applied to identity testing.

More recently another method has been developed to detect VNTR sequences which uses polymerase chain reaction (PCR) amplification technology (1). Similar to RFLP analysis (see Chapter 10), the PCR based approach first involves isolating a DNA sample. Next, rather than fragmenting the DNA with a restriction endonuclease, the PCR amplification technique is used to generate millions of copies of the specific segment of DNA containing the VNTR sequence (see *Figure 1*). The PCR-generated DNA fragments can be displayed very rapidly using acrylamide gel electrophoresis and a simple silver staining procedure. The approach for the detection of VNTRs using the PCR reaction is referred to as amplified fragment length polymorphism (AMPFLP) analysis. This chapter describes the analysis of the VNTR locus D1S80 by the AMPFLP method and how it is used to investigate cases of disputed parentage (2, 3).

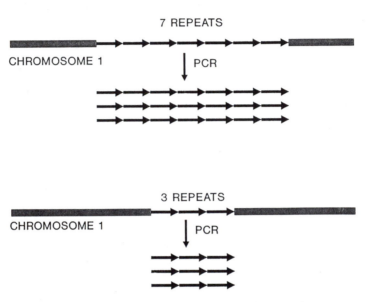

Figure 1. Amplification of a fragment length polymorphism on chromosome 1 at the D1S80 locus using the PCR. The arrows of 7 repeats and 3 repeats indicate the short DNA sequence that is repeated tandemly, producing the fragment length polymorphism and therefore different alleles.

The following equipment is required for AMPFLP analysis.

- Rotator (specimen tube rotator, Fischer Sci #13688-ID
- Microfuge
- Specimen preparation hoods/amplification set-up hoods (for example, Labconco tissue culture enclosure, Fischer Sci #16-105; see also Chapter 4)
- Temperature-controlled baths
- Hot plate
- 2-ml screw-cap centrifuge tubes
- Perkin–Elmer Cetus Gene Amp tubes or Continental Products 0.5-ml snap-cap tubes (#3435)

- Positive-displacement pipetters with appropriate tips or Rainin-type pipetters with ART filter tips from Continental Products (#2069 to fit P200, #2149 to fit P20)

- Thermal cycler (for example, Perkin–Elmer, Model 480)

- Vertical acrylamide gel apparatus (Hoefer, SE 410–24 cm)

- Electrophoresis power supply (capable of constant voltage to 250 volts, Hoefer PS250)

- Platform shaker (New Brunswick Sci. Co., Gyrator shaker, Model G-2, available from Fischer Scientific)

2. Preparation of DNA

There are a variety of protocols that can be used to prepare DNA from fresh whole blood or dried blood stains. The main objective is to lyse white blood cells releasing nuclear DNA and to remove inhibitors of the DNA polymerase used during the amplification process. It is important to prevent contamination from extraneous DNA and nucleases released from the individual preparing the samples. Precautions must be taken to avoid carry-over of previously amplified PCR products inadvertently introduced from reagents, equipment (especially pipettes), and other specimens. Autoclaved solutions and labware should be used whenever possible and gloves should be worn throughout the procedure. It is also a good idea to dedicate equipment and laboratory space for specimen preparation, physically segregating this process from the analysis of PCR products (see Chapter 4 for further details).

Protocol 1. Rapid preparation of DNA from fresh blood or dried stains using Chelex

1. Pipette 1 ml of sterile glass-distilled water into a 2-ml microcentrifuge tube. Add 5 µl of whole blood or 3 mm^2 of material with dried blood and mix gently.

2. Incubate at room temperature for 20 min with gentle mixing (a specimen tube rotator works well).

3. Centrifuge at 10 000 g for 3 min. Remove all but 20–30 µl of the supernatant, leaving the cell pellet or fabric substrate in the tube.

4. Add 5% Chelex solution to a final volume of 200 µl (Chelex 100–200 mesh, sodium form, biotechnology grade (Bio-Rad), 5% solution in sterile distilled water). When pipetting Chelex stock solutions, the resin beads must be distributed evenly in solution and the Chelex must be pipetted with a wide-bore pipette tip, such as the one that fits a P1000 pipette.

5. Incubate at 56°C for 30 min then vortex rapidly for 10 sec.

Protocol 1. *Continued*

6. Boil the sample for 8 min, vortex for 10 sec, then centrifuge at $10\,000\,g$ for 3 min.

7. Add $20\,\mu$l of the supernatant to each PCR reaction.

---

3. Amplification of the D1S80 locus

The PCR technique is an *in vitro* method of nucleic acid synthesis whereby a specific segment of DNA can be replicated (see Chapter 4). The specificity of the PCR process is controlled by designing two oligonucleotide primers that anneal only with the nucleotide sequences that flank the VNTR DNA selected for amplification. These primers hybridize to opposite strands of the target VNTR sequence and are orientated so that synthesis by DNA polymerase proceeds across the region between the primers.

A procedure for the amplification of the D1S80 locus is described in *Protocol 2*.

For paternity testing, blood samples from the mother, child, and alleged father, prepared for PCR by *Protocol 1*, are placed in a tube with the PCR master mix (see *Table 1*). Each tube is then placed into the thermal cycler. The PCR reaction components and thermal cycling times described in *Protocol 2* have been optimized for the Perkin–Elmer model 480 and may need adjusting if other thermal cyclers are used. During the first phase of the amplification process the thermal cycler raises the temperature of the PCR reaction to approximately 95°C to denature the double-stranded DNA molecule. The sample is then cooled to a temperature permissive for primer annealing to the sequences flanking the VNTR. In the last phase the temperature is raised to 72°C while *Taq* polymerase adds nucleotides to the 3′-end of each primer. The amplification PCR product is an exact copy of the target VNTR. Each series of denaturation, primer annealing, and DNA synthesis

---

Table 1. Master mix recipe for amplification of the D1S80 locus

2 × Master mix

- 2 × *Taq* polymerase buffer: 20 mM Tris–HCl, 100 mM KCl, 3 mM $MgCl_2$, 0.02% gelatin, pH 8.4
- 1.6 mM of each of dATP, dCTP, dGTP, and dTTP in sterile distilled water
- 50 pmol of each HPLC purified primer:
  MCT 118–1 5′ GGGAAACTGGCCTCCAAACACTGCCCGCCG 3′
  MCT 118–2 5′ GGGTCTTGTTGGAGATGCACGTGCCCCTTGC 3′
- 5 units of recombinant *Taq* polymerase per 50 μl of 2 × Master mix (added just prior to use)

---

reactions results in a doubling of the target DNA from the previous PCR cycle. Following analytical gel electrophoresis the PCR products from the mother, child, and alleged father can be visualized rapidly with a simple silver-staining technique.

Different gloves and lab coat from those used in sample preparation should be worn while preparing PCR reactions. It is strongly suggested that specimen preparation and amplification protocols be performed in separate enclosed hoods in a location separate from analysis of the PCR products. Positive displacement pipette with appropriate tips, or filter-top tips from Continental Products should be used at all steps of the amplification procedure.

Protocol 2. Amplification of the D1S80 locus

Materials
- Master mix (see *Table 1*)
- *Taq* polymerase (Perkin–Elmer Cetus #8010046)
- 5 × sucrose loading buffer: 60% (w/v) sucrose, 0.25% xylene cyanol, 0.25% bromophenol blue

Method
- Turn on the thermal cycler 30 min before use
- 10 min before use, pre-warm the thermal cycler to 95°C

1. Add 30 µl of sterile distilled water to each amplification tube. For each thermal cycler run, include a known DNA control and a negative water control (see also Chapter 4).

2. Add 20 µl of DNA that has been prepared by *Protocol 1* or 300 ng of purified DNA (in 20 µl) to each amplification tube except the negative control. Add 20 µl of sterile distilled water to the negative control.

3. Thaw 2 × master mix (see *Table 1*), add 5 Units of *Taq* polymerase per 50 µl, and mix gently by inversion.

4. Add 50 µl of 2 × master mix to each DNA sample and the positive and negative controls.

5. Centrifuge each tube for 5 sec at 12 000 g in a microfuge and place in the thermal cycler.

6. Heat the samples initially to 95°C for 5 min then perform a total of 25 amplification cycles, with each cycle consisting of 1 min at 65°C for primer annealing, 1 min at 72°C for primer extension and 1 min at 95°C for denaturation. The last cycle is followed by slow cooling to 50°C at a rate of 1°C per 30 sec and then rapid cooling to 4°C. The sample should then be held at 4°C until they are removed from the cycler.

Protocol 2. *Continued*

7. Centrifuge the tubes for 5 sec at 12 000 *g* in a microfuge and store them at −20°C until the PCR products are analysed.

8. To a tube containing 5 μl of 5 × sucrose loading buffer add 20 μl of the PCR reaction mix.

9. For the Child/Alleged Father Mix, add 10 μl of each to a tube containing 5 μl of 5 × sucrose loading buffer.

---

4. Detection of PCR products

The abundance of the PCR products allows visualization of fragment length polymorphisms directly, eliminating the need for Southern transfer and hybridization with a radiolabelled probe. The interpretation of AMPFLP results is simple, quick and easy to standardize. There is no need for complex DNA-sizing algorithms as in RFLP analysis. Instead, a composite mixture of all possible alleles is processed along with the samples. Each individual's genetic profile is determined rapidly and reproducibly by comparison with the composite allele standards.

The amplified products from the D1S80 locus range in size from approximately 375 base-pairs to 850 base-pairs. The repeat length is only 16 base-pairs long. For good resolution and separation of alleles that may differ by only one repeat length we use a 0.75 mm × 24 cm, 6% polyacrylamide gel. Pour a gel according to the recipe in *Table 2*: gels should be poured on Gel Bond, available from FMC, according to the manufacturer's specifications. Gel Bond is a solid support that allows for easier handling of the gel during the staining process. Let the gel polymerize for at least 1.0–1.5 h, then remove the comb and rinse the wells three times with 1 × TBE (see *Table 2*). Load the wells dry with 15–20 μl of PCR product and then overlay each well slowly with 1 × TBE before putting on the top of the gel apparatus. The specimens for each case are flanked with an appropriate allele ladder made up in 1 × *Taq* buffer (see *Table 1*).

For example:

Lane 1	2	3	4	5	6
Allele Mix	Mother	Child	Alleged Father	Child/Alleged Father Mix	Allele Mix

Electrophoretic time for the analysis of D1S80 PCR products is approximately 1800 volt-hours. Following electrophoresis, the gel is silver-stained according to *Protocol 3*. Specimens are usually prepared in the morning, amplified in the afternoon, subjected to electrophoresis at 110 constant volts overnight and then stained the next day.

Table 2. 6% polyacrylamide gel recipe (14 cm × 24 cm × 0.75 mm)

• 30% acrylamide (29.1 polyacrylamide: 0.9 bis-acrylamide)	7.5 ml
• 20% glycerol	13.12 ml
• 10% ammonium persulfate	0.375 ml
• 10 × Tris–borate–EDTA (TBE): 0.45 M Tris–HCl, 0.025 M sodium EDTA, 0.9 M boric acid, pH 9.0	3.75 ml
• Glass-distilled water	12.7 ml
• TEMED	0.0375 ml
Total volume	37.5 ml

Protocol 3. Silver-staining of polyacrylamide gels

Materials
- 1% (v/v) nitric acid
- 0.012 M silver nitrate
- 37% formaldehyde solution (see Chapter 1)
- 10% (v/v) acetic acid
- Developer solution: 0.28 M sodium carbonate

Method
1. Disassemble the gel and place it (still attached to Gel Bond), gel-side up in a glass tray. Use approximately 350 ml of each solution. During each step in the staining process, the gel should be subjected to vigorous shaking on a platform shaker. If the gel has not been poured on Gel Bond, do not shake it vigorously.
2. Treat the gel for 10 min in 10% ethanol and pour off.
3. Treat the gel in 1% nitric acid for 6 min and pour off the solution.
4. Rinse once briefly with distilled water followed by a 5 min rinse in water with shaking.
5. Treat for 40 min in 0.012 M silver nitrate and pour off the solution.
6. Rinse twice in distilled water with shaking.
7. Add 175 µl of formaldehyde to 350 ml of developer solution that has been chilled to 4°C. Add developer to the gel while it is still cold, shake the gel until the PCR products are visualized and pour off the developer.
8. Stop the development by shaking the gel in 10% (v/v) acetic acid for 5 min and pour off the solution.

Protocol 3. *Continued*

9. Rinse the gel in distilled water for at least 5 min.

10. The gel can be air-dried (if poured on Gel Bond), dried on a vacuum drier on to Whatman filter paper, or covered with porous cellophane (available from Hoefer Scientific) and then dried on a vacuum drier.

---

5. Interpretation of results

The allele designations for each specimen can be assigned directly by matching the electrophoretic position of each silver-stained band in the mother, child, and alleged father with a corresponding band present in the allele ladder. For example, in *Figure 2*, the mother's DNA that is part of the paternity case designated 'inclusion' exhibits alleles 6 and 9. The child also possesses alleles 6 and 9, while the alleged father has alleles 3 and 9. Since the child may have inherited either the 6 or 9 allele from its mother, the biological father must possess either a 6 or a 9 allele. The alleged father does possess a number 9 allele and, therefore, the result is consistent with paternity. The probability that the alleged father actually produced the sperm that conceived this child is related to the frequency of the number 9 allele in his ethnic group (see *Table 3*).

In the paternity case designated as an 'exclusion', the mother is homozygous for allele 11. The child possesses alleles 3 and 11, while the alleged father has alleles 9 and 13. The child has inherited allele 11 from its mother and therefore, the biological father must possess an allele 9. Since the alleged

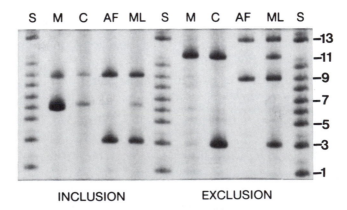

INCLUSION   EXCLUSION

Figure 2. AMPFLP-based paternity testing. A silver-stained polyacrylamide gel displaying D1S80 amplified product from two paternity cases. The position of each PCR band in the mother (M), child (C), and alleged father (AF) is compared with its corresponding allele in the composite allele mixture lane (S).

Table 3. Allele frequencies for locus D1S80

Allele	Caucasians	Blacks	Hispanics
1	0.007	0.0002	0.01
2	0	0.041	0.007
3	0.209	0.068	0.296
4	0.004	0	0.007
5	0.022	0.023	0.022
6	0.029	0.166	0.035
7	0.02	0.072	0.054
8	0.009	0.035	0.012
9	0.321	0.188	0.296
10	0.042	0.052	0.077
11	0.016	0.014	0.03
12	0.007	0.004	0.027
13	0.038	0.135	0.069
14	0.056	0.046	0.067
15	0.02	0.014	0.035
16	0.08	0.056	0.052
17	0.007	0.004	0
18	0.007	0.068	0.01
19	0	0.004	0
20	0.002	0.004	0.002
21	0.002	0	0
22	0.002	0	0
23	0.002	0.002	0

father does not possess allele 9, this result is not consistent with paternity; that is, the alleged father is excluded from paternity with this test.

Accurate calculations for probability of paternity or identity require good estimates of allele or genotype frequencies. The allele frequencies presented in *Table 3* for D1S80 were obtained from a random sampling of blood specimens originating from over 500 sites in the United States. The frequencies are based on 400-plus observed alleles from each of the Caucasian, North American Black and Hispanic ethnic groups. The average power of exclusion for D1S80 is 0.6028 for Caucasians, 0.7837 for Blacks and 0.7250 for Hispanics.

Generally, four or five independently segregating loci should be analysed by PCR-based AMPFLP techniques to offer satisfactory laboratory evidence regarding parentage, that is, an exclusion or a very high probability of paternity or identity match (4, 5). The principles and protocols presented in this chapter can be applied equally to the establishment of identity in forensic pathology.

References

1. Mullis, K. and Faloona, F. (1987). In *Methods in enzymology,* Vol. 155 (ed. R. Wu), pp. 335–50. Academic Press, London.
2. Nakamura, Y., Carlson, M., Krapcho, K., and White, R. (1988). *Nucleic Acids Res.,* **16,** 9364,
3. Budowle, B., Chakraborty, R., Giusti, A. M., Eisenberg, A. J., and Allen, R. C. (1991). *Am. J. Hum. Genet.,* **48,** 926.
4. American Association of Blood Banks, Virginia. (1983). *Inclusion probabilities in parentage testing* (ed. R. Walker).
5. Kirby, L. T. (1990). In *DNA fingerprinting,* pp. 149–78. Stockton Press, New York.

10

DNA analysis of archival material and its application to tumour pathology

W. YASUI, H. ITO, and E. TAHARA

1. Introduction

Recent progress in molecular biology has introduced new concepts in the understanding of the pathogenesis of disease. For example, recent evidence indicates that the development and progression of cancer requires multiple genetic alterations affecting both oncogenes and tumour suppressor genes (anti-oncogenes) (1–3). These advances are largely attributable to new molecular techniques which allow rapid and simple analysis of genetic alterations in tumour biopsies.

This chapter describes the application of molecular techniques to the analysis of tumours, with emphasis on techniques applicable to routine archival material. It begins with protocols for the detection of point mutations in both fresh and archival clinical specimens. This is followed by the detection of gene amplification and, in the latter part, the detection of loss of heterozygosity (LOH) in tumours, using restriction fragment length polymorphism (RFLP) analysis.

2. The detection of point mutations

At least three methods are available for the detection of point mutations in clinical samples. The first is the allele-specific oligonucleotide hybridization method which is based on the principle that a duplex of oligonucleotides with even a single base mismatch is extremely unstable (4). This method involves two steps:

(a) amplification of DNA from the sample by the polymerase chain reaction (PCR);

(b) screening the amplified DNA on filters with radiolabelled synthetic oligomers whose sequences correspond to possible amino acid substitutions.

The second method is direct sequencing of DNA amplified by PCR (see Chapter 5). The other is a rapid and sensitive method called single-strand conformation polymorphism (SSCP) analysis, which has been developed recently (5). This method is based on the principle that specific regions of genomic DNA can be labelled and amplified simultaneously by using labelled substrates in the PCR and that the mobility of single-stranded nucleic acid in non-denaturing polyacrylamide gels depends not only on its size but also on its sequence. These methods are discussed in turn below.

2.1 Allele-specific oligonucleotide hybridization

This involves initial DNA extraction from clinical samples and DNA amplification by the PCR (6).

2.1.1 DNA extraction

DNA can be extracted from formalin-fixed and paraffin-embedded tissue prepared for routine histopathological examination (see *Protocol 1*) (7, 8). Formylation of nucleic acids produces Schiff bases on free amino groups of nucleotides, and exposure of nucleo-proteins to formaldehyde results in the formation of crosslinks between proteins and DNA (9). These processes are reversible in aqueous solution, implying that DNA can be recovered from formalin-fixed tissue. Although DNA extracted in this way is not completely intact, it is double-stranded, cleavable with restriction endonucleases, can be hybridized with labelled probes, and amplified by using the PCR. If available, DNA extracted from fresh or frozen tissues is more suitable for a variety of standard molecular techniques (10) and protocols for this extraction are given in Chapter 1.

Protocol 1. Extraction of DNA from formalin-fixed and paraffin-embedded tissue specimens

Materials
- 10 × SSC buffer: 1.5 M NaCl, 0.15 M sodium citrate; adjust the pH to 7.0 with 1 M HCl
- Proteinase K (10 mg/ml in sterile distilled water)
- Phenol:chloroform:isoamylalcohol (25:24:1)
- 3 M sodium acetate, pH 5.2
- Ice-cold absolute ethanol
- TE buffer: 10 mM Tris–HCl, pH 8.0, 1 mM EDTA

Method
1. Cut five to ten 20 μm sections on to glass slides using a microtome.[a]

2. Deparaffinize as for histology sections by sequential incubation for 10 min in three changes each of xylene, 100% ethanol, 70% ethanol, and water. Finally, rinse with 1 × SSC.

3. Cut out part of interest from the sections, using a sterile scalpel blade.

4. Incubate each specimen in 3 ml of digestion solution (2470 μl of 1 × SSC, 500 μl of 10% SDS, and 30 μl of proteinase K (10 mg/ml)) at 37°C for five to ten days. Proteinase K should be added every two or three days.

5. Extract with phenol:chloroform:isoamylalcohol (25:24:1) (see Chapter 1).

6. Precipitate with ice-cold ethanol and redissolve the pellet in 200 μl of TE buffer (see Chapter 1).

[a] The blocks should not be used if autolysis or necrosis is evident. The most important cause of failure to recover nearly intact DNA is the use of inadequately fixed tissue.

2.1.2 Amplification of DNA by PCR

The basic PCR has been described at length in Chapter 4. The following section discusses a protocol used prior to allele-specific oligonucleotide hybridization in human tumours. There are no rules that guarantee the choice of an effective primer pair. However, the following guide-lines are helpful:

- Primers with an average G + C content of around 50% and a random base distribution should be selected.

- Sequences with significant secondary structure should be avoided. Computer programs such as Squiggles or Circles are useful for revealing these structures.

- The primers should be checked against each other for complementarity. Avoiding primers with 3′ overlaps will reduce the incidence of 'primer–dimer' artefacts.

Protocol 2. Amplification of genomic DNA by PCR

Materials
- 10 × PCR buffer: 500 mM KCl, 100 mM Tris–HCl, pH 8.3, 15 mM $MgCl_2$, 0.1% (w/v) gelatin
- 10 × dNTP mixture: 2 mM each of dATP, dCTP, dTTP, and dGTP
- *Taq* polymerase (5000 units/ml, Perkin–Elmer Cetus)
- 250–500 ng of genomic DNA
- 10 × Primer 1: 10 μM in TE buffer
- 10 × Primer 2: 10 μM in TE buffer

Protocol 2. *Continued*

Method

1. Add 10 μl each of 10 × PCR buffer, 10 × dNTP mixture, 10 × primer 1, 10 × primer 2, DNA samples and distilled water to give a total volume of 100 μl in sterile Eppendorf tubes.

2. Heat the reaction mixture at 90–95°C for 10 min to denature the DNA.

3. Add 2.5 units (0.5 μl) of *Taq* polymerase, mix gently and overlay with a few drops of mineral oil.

4. Amplify for 30 to 35 cycles.[a] (See Chapter 4.)

5. Check the amplified fragments on ethidium-bromide-stained agarose gels using a part of the reactants (see Chapters 1 and 4).

[a] DNA from de-waxed formalin fixed tissues may require more cycles (up to 50), probably as a result of de-purination of DNA by formalin.

2.1.3 Dot blot hybridization

The second step of allele-specific oligonucleotide hybridization is DNA immobilization on to filters and hybridization with an oligonucleotide which spans the mutation of interest. The oligonucleotide is usually terminally labelled with a radiolabelled nucleotide (see Chapter 2) (11, 12). The oligonucleotide hybridization method is given in *Protocol 3*.

Protocol 3. Dot blot hybridization

A. DNA immobilization on to nitrocellulose filters

Materials

- DNA samples
- 1 M NaOH
- TE buffer (see *Protocol 1*)
- 2 M ammonium acetate
- Nitrocellulose filters (BA85, Schleicher & Schuell)
- Dot blot apparatus (for example, Manifold-II, Schleicher & Schuell; Bio-Dot SF, Bio-Rad)

Method

1. Dissolve 10 μg of DNA in 50 μl 0.4 M NaOH in TE buffer.

2. Incubate for 10 min at 37°C, and then treat with an equal volume of 2 M ammonium acetate.

3. Apply the samples to nitrocellulose using the dot-blot apparatus or by hand (see Chapter 3).

4. Bake the filter at 80°C for 2 h under vacuum.

B. Hybridization procedure

Materials

- 20 × SSPE: 3 M NaCl, 200 mM NaH_2PO_4, 50 mM EDTA, pH 7.4
- 100 × Denhardt's solution: 2% (w/v) Ficoll, 2% (w/v) polyvinylpyrrolidone, and 2% (w/v) bovine serum albumin (Fraction V) in sterile distilled water
- 10 × SSC (see *Protocol 1*)
- Washing buffer: 3 M tetramethyl-ammonium chloride, 50 mM Tris–HCl, pH 8.0, 2 mM EDTA, 0.1% SDS
- 10% SDS
- ^{32}P-end-labelled probe[a]

Method

1. Pre-hybridize for 2 h at 65°C in pre-hybridization solution: 5 × SSPE, 5 × Denhardt's solution, and 5% SDS.

2. Add the end-labelled probe (1–1.5 µM) to the pre-hybridization solution.

3. Hybridize for 6 h to overnight at the same temperature as pre-hybridization.

4. Wash the filter in a tray with shaking as follows:

    (a) 5 × SSC, room temperature (two changes each of 15 min)

    (b) Washing buffer, 61°C (once for 20 min)

    (c) 5 × SSC, room temperature (once for 10 min)

5. Allow the filter to air-dry and expose to X-ray film (see Chapter 3).

---

[a] The end-labelling procedure is described in Chapter 2 and refs. 10 and 14.

A representative result obtained using the above procedures is shown in *Figure 1*. The stability and efficiency of hybridization depends on the sequence as well as length of the probe (oligonucleotide) used. Therefore, optimal washing conditions should be chosen by experiment for each probe. Mutated and non-mutated control DNA should be included in each experiment, and hybridization carried out with probes for both the wild-type and mutated sequences.

2.2 Direct sequencing of DNA amplified by PCR

In vitro amplification of DNA using PCR and subsequent direct sequencing of

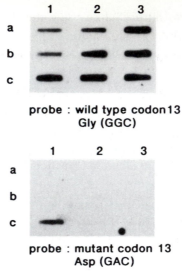

probe : wild type codon 13
Gly (GGC)

probe : mutant codon 13
Asp (GAC)

Figure 1. Detection of Ki-*ras* point mutation in human gastric carcinomas by using allelic specific oligonucleotide hybridization combined with PCR. 10 μg of DNA amplified *in vitro* using the primers (5′–TTTTTATTATAAGGCCTGCT–3′; 5′–GTCCACAAAATGATTCTG AA–3′) (13) were applied to dot blotting. Hybridization was performed with ^{32}P-end-labelled probe of wild-type Ki-*ras* and the Asp-mutant at codon 13.

the product is an alternative approach to the detection of mutations (4, 14). Sequencing of PCR products is described in Chapter 5.

2.3 SSCP analysis

In this method, the sequences of interest are amplified and labelled by PCR using labelled primers or a labelled nucleotide (5). The procedure is described in *Protocol 4*. Most single-base changes in fragments of up to 200 bases can be detected as mobility shifts by SSCP analysis (for example, *ras* point mutation) (15, 16). However, this mobility shift cannot determine either the precise position or the exact nature of the base changes. Nevertheless, SSCP analysis is useful for screening for point mutations as it is simple, fast, and efficient. When mutated cases are identified, other methods such as direct sequencing must be used to determine the base substitutions.

Protocol 4. Single-strand conformation polymorphism (SSCP) analysis

Materials

• DNA sample (50 ng/μl)

• 10 × PK buffer: 500 mM Tris–HCl, pH 8.3, 100 mM MgCl$_2$, 50 mM DTT

- Polynucleotide kinase (10 units/μl)
- γ-^{32}P-ATP (160 mCi/ml, 700 Ci/mmol)
- 10 × PCR buffer: 200 mM Tris–HCl, pH 8.3, 500 mM KCl, 20 mM MgCl$_2$, 1 mg/ml BSA
- dNTP mixture: 1.25 mM of each of dATP, dCTP, dGTP, and dTTP
- *Taq* polymerase (5000 units/ml, Perkin–Elmer Cetus)
- A pair of primers each at 10 μM
- Formamide dye: 95% formamide, 10 mM EDTA, 0.05% bromophenol blue, 0.05% xylene cyanol
- Acrylamide solution: 49% acrylamide, 1% *N,N'*-methylene-bis-acrylamide
- 10 × TBE buffer: 0.5 M Tris base, 0.5 M boric acid, 20 mM EDTA
- DNA thermal cycler (for example, Cetus thermal cycler; Pharmacia GeneA-TAQ)
- Sequencing gel apparatus

Method

1. Mix the following solutions and incubate at 37°C for 30 min.
   - Water .... 1.0 μl
   - Primers .... 1.0 μl of each
   - 10 × PK buffer .... 0.5 μl
   - γ-^{32}P-ATP .... 1 μl
   - Polynucleotide kinase ... 0.5 μl

2. Add 255 μl of H$_2$O, 40 μl of 10 × PCR buffer and 20 μl of dNTP mixture.

3. Take 80 μl of the mixture from step 2, and add 0.5 μl of *Taq* polymerase.

4. Transfer 4 μl of the mixture from step 3 to a 0.5-ml tube, add 1 μl of DNA sample, mix, and overlay with a few drops of mineral oil.

5. Heat at 94°C for 1 min and start amplification for 30 cycles (94°C for 20 sec to denature and 60°C or 65°C for 2 min to anneal and elongate the chain).

6. Add 45 μl of formamide dye, mix, and centrifuge at 12000 g for 2 min.

7. Set up the apparatus for polyacrylamide gel electrophoresis. The recipe is as follows:
   - 10 × TBE buffer .... 1.5 ml
   - 50% glycerol .... 3.0 ml
   - Acrylamide solution ..... 3.0 ml

Protocol 4. *Continued*

- 1.6% ammonium persulfate   1.0 ml
- Water   21.5 ml
- TEMED   30 μl

8. Heat the PCR product in formamide dye at 80°C for 5 min, and apply 1 μl of the samples per lane.

9. Start electrophoresis at a constant power of 40 W while cooling, and stop electrophoresis when the xylene cyanol has migrated to 5 cm from the bottom of the gel.

10. Dry the gel and expose to X-ray film (see Chapter 3).

---

3. The detection of gene amplification

Gene amplification can be detected in formalin-fixed paraffin-embedded tissues by dot-blot analysis (see *Protocol 5*). In this case, the specificity of hybridization should be carefully checked. On fresh or frozen samples, Southern blot analysis (see Chapter 3) is preferable (10, 17). Amplification is determined by comparing the signal intensity by densitometry of either equal parallel samples hybridized with a single-copy gene (such as β-globin) and the gene of interest, or by probing and counting the filter sequentially with these two probes.

Protocol 5. Detection of gene amplification by dot-blot hybridization

Materials

- DNA samples (from formalin-fixed and paraffin embedded tissue: see *Protocol 1*)
- TE/NaOH buffer: 0.4 M NaOH in TE buffer (see *Protocol 1*)
- Salmon sperm DNA (sonicated)
- 20 × SSPE (see *Protocol 1*)
- 100 × Denhardt's solution (see *Protocol 3*)
- 10 × SSC (see *Protocol 1*)
- Pre-hybridization solution (*see Protocol 3*)
- ^{32}P-end-labelled or random-primed ^{32}P-labelled probe[a]
- Nitrocellulose filter
- Dot-blot apparatus

Method

1. Extract DNA from formalin-fixed paraffin-embedded tissues as in *Protocol 1*.

2. Dissolve 10 μg of DNA in TE/NaOH buffer to make 50 μl of solution.

3. Make serial dilutions of the DNA solution of interest in TE/NaOH buffer (for example, ×1/2, ×1/4, ×1/8, ×1/16). Add salmon sperm DNA in the same buffer to give a final DNA concentration of 10 μg per 50 μl.

4. Incubate for 10 min at 37°C, and then add 50 μl of 2 M ammonium acetate.

5. Transfer the sample on to nitrocellulose filter using either a dot blot apparatus or by hand (see Chapter 4), and bake the filter at 80°C for 2 h under vacuum.

6. Pre-hybridize and hybridize the filter with a ^{32}P-labelled probe as in *Protocol 3*.

7. Wash the filter in a tray with shaking as follows:
   (a) 0.1 × SSC, 0.1% SDS, room temperature (two changes each of 15 min).
   (b) 0.1 × SSC, 0.1% SDS, 65°C (two changes each of 30 min).
   (c) 0.1 × SSC, room temperature (two changes each of 10 min).

8. Allow the filter to air-dry and expose it to X-ray film.

[a] The labelling of probes with ^{32}P is described in Chapter 2.

---

4. Detection of loss of heterozygosity (LOH)

The principle of RFLP analysis is shown in *Figure 2* (18, 19). DNA restriction enzymes recognize specific sequences in DNA and, by catalysing endonucleolytic cleavages, create fragments of certain lengths which are displayed by agarose gel electrophoresis. Differences among individuals in the lengths of a particular restriction fragment result from many kinds of genetic differences. One or more individual bases may differ, resulting in either loss or formation of cleavage sites. Insertion or deletion of blocks of DNA may also alter the size of the fragments. In both cases, the altered mobility of restriction fragments on the gel can readily be detected by Southern blotting and hybridization with a labelled DNA probe which recognizes an adjacent or 'linked' DNA sequence. This method has been applied to genetic linkage analysis for genetic disease as well as to the detection of specific gene deletions in tumours, and can be applied to any extracted DNA, whether from fresh or archival material.

4.1 Choice of RFLP markers and restriction enzymes

The frequency of heterozygosity among individuals in the population is important in the selection of useful markers. As alleles at genetic loci are

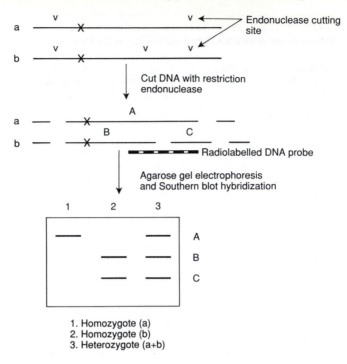

1. Homozygote (a)
2. Homozygote (b)
3. Heterozygote (a+b)

Figure 2. Detection of RFLP by Southern blot hybridization.

distributed essentially by chance among humans, the expected frequency of heterozygosity at a marker locus (and hence its informativeness) is directly related to the number of common alleles present in the population. The more alleles a locus exhibits, the more likely it is that an individual will be heterozygous. The polymorphic information content (PIC) is calculated from the allele frequencies as described in the HGM9 DNA Committee Report (18, 20). The loci with PIC greater than 0.4 are the multi-allelic systems, and most are variable number of tandem repeats (VNTRs) although some are haplotypes (21). Many VNTR sequences have been cloned, and are available as RFLP markers (22). The choice of restriction endonuclease is also important. *Msp*I and *Taq*I give high resolution of RFLPs (see *Table 1*). Other enzymes such as *Pst*I and *Bgl*II also recognize RFLPs frequently. The enzyme which gives the best resolution should be used, and more than two may be required in a given situation.

4.2 RFLP analysis

Conventional RFLP analysis has two components:

(a) restriction endonuclease digestion and agarose gel electrophoresis, and

(b) Southern blotting and hybridization.

Table 1. Recognition sequences and reaction conditions of restriction endonucleases frequently used in RFLP analysis[a]

Restriction endonuclease	Recognition sequence	NaCl concentration (mM)	Reaction temperature (°C)
*Bam*HI	G:GATC	150	37
Bgl II	A:GATCT	100	37
*Eco*RI	G:AATTC	50	37
*Hin*dIII	A:AGCTT	50	37
*Msp*I	C:CGG	0	37
*Pst*I	CTGCA:G	100	37
*Pvu*II	CAG:CTG	50	37
*Taq*I	T:CGA	100	65

[a] Compiled from the catalogues of New England Biolabs (1988/1989) and Bethesda Research Laboratories (1990).

The former is described in *Protocol 6*. Southern transfer and filter hybridization are described in detail in Chapter 3. A representative result obtained by RFLP analysis is shown in *Figure 3*. Nylon filters can be used instead of nitrocellulose (10, 14) and the transfer protocol for these is faster and simpler than that for nitrocellulose filters. In addition, nylon filters are more robust than nitrocellulose and can be reused for several hybridization reactions. An

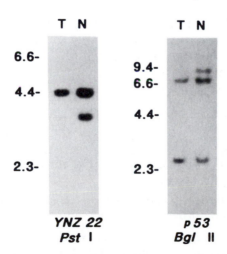

Figure 3. Detection of LOH on chromosome 17p in tumour tissue by RFLP analysis. 10 μg of DNA fragments digested with *Pst*I or *Bgl* II was subjected to 0.8% agarose gel electrophoresis, followed by Southern transfer. Hybridization was carried out with random primed ³²P-labelled probes as indicated (YNZ22 = a VNTR marker; p53 = a tumour suppressor gene. Both sequences are located on the short arm of chromosome 17). T = gastric cancer tissue, N = non-neoplastic gastric mucosa.

alternative approach to RFLP analysis can be employed if the sequence of the region of the polymorphism is known. By designing appropriate primers, the region containing the polymorphism can be amplified by PCR and then cleaved with the appropriate restriction endonuclease: the loss or gain of a restriction site can be determined by the absence or presence respectively of a cleaved PCR product. This approach is similarly applicable to archival material.

Protocol 6. Restriction endonuclease digestion and agarose gel electrophoresis

Materials

- 10 × restriction endonuclease buffer: 100 mM Tris–HCl, pH 7.5, 100 mM MgCl$_2$, 10 mM dithiothreitol, and 0, 0.5, 1.0, or 1.5 M NaCl[a]
- Restriction endonuclease (see *Table 1*)
- 10 × loading buffer: 50% glycerol, 0.25% bromophenol blue, 0.25% xylene cyanol, 100 mM EDTA, pH 8.0)
- 10 × TBE buffer: 0.89 µM Tris base, 0.89 M boric acid, 20 mM EDTA—adjust the pH to 8.1–8.2 with solid boric acid
- Electrophoresis-grade agarose
- TE buffer (see *Protocol 1*)
- Ethidium bromide solution: 1000 × stock solution, 0.5 mg/ml, protect from light
- Horizontal gel electrophoresis apparatus

Method

1. Pipette the following into a microcentrifuge tube: 20 µl of 10 × restriction endonuclease buffer, x µl of DNA (30 µg), and sterile distilled water to give a total volume of 180 µl.[b]

2. Add endonuclease (1 to 5 U/µg DNA). The volume added should be less than 1/10th of the final mixture.

3. Incubate the mixture overnight at the recommended temperature (see *Table 1*).

4. Extract the DNA fragments from the reaction mixture with phenol/chloroform and precipitate in ethanol (see Chapter 1).

5. Suspend the precipitated DNA in TE buffer and measure the DNA concentration (see Chapter 1).

6. Combine 10 µg of DNA samples, 3 µl of 10 × loading buffer, and TE buffer to a total volume of 30 µl.

7. Load the whole sample on to an agarose gel and run the electrophoresis in 0.5 × TBE buffer.[c]

8. Stain the gel in ethidium bromide and observe on a UV light source (see Chapter 1).

[a] The concentration of NaCl depends on the restriction endonuclease (see *Table 1*).
[b] The amount of DNA to be cleaved or the reaction volume can be changed provided the proportion of the components remain constant.
[c] Appropriate concentrations of agarose should be chosen (see *Table 2*).

Table 2. Appropriate agarose concentrations for separating DNA fragments of various sizes

Agarose (%)	Linear DNA fragments (kb)	Restriction endonuclease
0.8	20 to 10	*Bam*HI, *Bgl* II, *Eco*RI *Pst*I, *Pvu*II
1.0	10 to 0.5	*Msp*I, *Taq*I
1.2	7 to 0.4	
1.5	4 to 0.2	

References

1. Tahara, E. (1990). *J. Cancer Res. Clin. Oncol.*, **116**, 121.
2. Aronson, S. A. and Tronick, S. R. (1985). In *Important advances in oncology* (ed. V. T. Devita). J. B. Lippincott, Philadelphia.
3. Stanbridge, E. J. (1990). *Science*, **247**, 12.
4. Innis, M. A., Gelfand, G. H., Sninsky, J. J., and White, T. J. (ed.) (1990). *PCR protocols, a guide to methods and applications*. Academic Press, San Diego, California.
5. Orita, M., Suzuki, Y., Sekiya, T., and Hayashi, K. (1989). *Genomics*, **5**, 874.
6. Impraim, C. C., Saiki, R. K., Erlich, H. A., and Teplitz, R. L. (1987). *Biochem. Biophys. Res. Commun.*, **142**, 710.
7. Goelz, S. E., Hamilton, S. R., and Vogelstein, B. (1985). *Biochem. Biophys. Res. Commun.*, **130**, 118.
8. Dubeau, L., Chandler, L. A., Gralow, J. R., Nichols, P. W., and Jones, P. A. (1986). *Cancer Res.*, **46**, 2964.
9. Jackson, V. (1978). *Cell*, **15**, 945.
10. Sambrook, J., Fritsch, E. F., and Maniatis, T. (ed.) (1989). *Molecular cloning, a laboratory manual* (2nd edn). Cold Spring Harbor Press, Cold Spring Harbor, NY.
11. Kafatos, F. C., Jones, C. W., and Efstratiadis, A. (1979). *Nucleic Acids Res.*, **7**, 1541.

12. White, B. A. and Bancroft, F. C. (1982). *J. Biol. Chem.,* **257,** 8569.
13. Janssen, J. W. G., Lyons, J., Steenvoorden, A. C. M., Seliger, H., and Bartram, C. R. (1987). *Nucleic Acids Res.,* **15,** 5669.
14. Ausubel, F. M., Brent, R., Kingston, R. E., Moore, D. D., Seidman, J. G., Smith, J. A., and Struhl, K. (ed.) (1987). *Current protocols in molecular biology.* John Wiley, New York.
15. Kanazawa, H., Noumi, T., and Futai, M. (1986). In *Methods in enzymology*, (ed. S. Fleischer and B. Fleischer), Vol. 126, p. 595. Academic Press, London.
16. Orita, M., Iwahana, H., Kanazawa, H., Hayashi, K., and Sekiya, T. (1989). *Proc. Natl. Acad. Sci. USA,* **86,** 2766.
17. Southern, E. M. (1975). *J. Mol. Biol.,* **98,** 503.
18. Botstein, D., White, R. L., Skolnick, M., and Davis, R. W. (1980). *Am. J. Hum. Genet.,* **32,** 314.
19. Alberts, B., Bray, D., Lewis, J., Raff, M., Roberts, K., and Watson, J. D. (ed.) (1989). *Molecular biology of the cell* (2nd edn). Garland Publishing, New York.
20. Pearson, P. L. (1987). *Cytogenet. Cell Genet.,* **46,** 390.
21. Kidd, K. K., Bowcock, A. M., Schmidtke, J., Track, R. K., Ricciuti, F., Hutchings, G., *et al.* (1989). *Cytogenet. Cell Genet.,* **51,** 622.
22. Nakamura, Y., Leppert, M., O'Connell, P., Wolff, R., Holm, T., Culver, M., *et al.* (1987). *Science*, **235,** 1616.

A1

Suppliers of specialist items

American Type Culture Collection, 12301 Parklawn Drive, Rockville, MD 20852, USA.

Amersham International plc, Lincoln Place, Green End, Aylesbury, Bucks HP20 2TP, UK.

Becton Dickinson, 1 Becton Drive, Franklin Lakes, New Jersey 07417-1880, USA. *Also:* Between Towns Road, Cowley, Oxford OX4 3ZY, UK.

Bethesda Research Laboratories, 8400 Helgerman Dourt, Gaithesburg, MD 20877, USA.

Bio 101 Inc., La Jolla, CA 92038-2284, USA.

Bio-Dyne B, Pall BioSupport, East Hills, NY, USA.

Bio-Rad Laboratories Ltd., Bio-Rad House, Marylands Avenue, Hemel Hempstead, Herts HP2 7TD, UK. *Also:* 3300 Regatta Drive, Richmond, California 94804, USA.

Boehringer Mannheim UK Ltd, Bell Lane, Lewes, East Sussex BN7 1LG, UK. *Also:* Boehringer Mannheim, PO Box 50414, Indianapolis, IN 46250, USA.

Cambridge Research Biochemicals, Gadbrook Park, Northwich, Cheshire CW9 7RA, UK. *Also:* Cambridge Research Biochemicals Inc., Wilmington, DE 19897, USA.

Continental Laboratory Products, 10171 Pacific Mesa Blvd, Suite 307, San Diego, CA 92121, USA.

CP Laboratories, PO Box 22, Bishop's Stortford, Herts CM23 3DH, UK.

Dako Ltd., 16 Manor Courtyard, Hughenden Avenue, High Wycombe, Bucks HP13 5RE, UK. *Also:* Dako Corporation, 6392 Via real, Carpinteria, CA 93013, USA.

Dynal, The Wirral, UK.

E.I. DuPont De Nemours and Co Inc., Market Street, 1007 Wilmington, Delaware 19898, USA. *Also:* DuPont UK Ltd., Wedgewood Way, Stevenage, Herts SG1 4QN, UK

FMC-Bioproducts, 5 Maple Street, Rockland, Maine 04841, USA. *Also:* FMC-pBioproducts (Europe), Risingevej, DK-2665 Vallensbaek Strand, Denmark.

Gelman Sciences Ltd., 10 Harrowden Road, Brackmills, Northampton NN4 0EZ, UK.

Gibco BRL, PO Box 35, Trident House, Renfrew Road, Paisley PA3 4EF, Scotland.

Hoefer Scientific Instruments, 654 Minnesota Street, San Francisco, CA 94107, USA.

ILS Ltd, 761 Main Avenue, Norwalk, CT 06859, USA.

International Medical Products, Zutphen, The Netherlands.

Life Technologies Inc., 8400 Helgerman Dourt, Gaithersburg, MD 20877, USA.

LTI Ltd., PO Box 35, Trident House, Renfrew Road, Paisley PA3 4EF, Scotland.

M. J. (Patterson) Scientific Ltd., Unit 2, Brookside, Colne Way, Watford, Herts WD2 4QJ

Nalgene, PO Box 20365, Rochester, NY 14602-0365, USA.

New England BioLabs Inc., 32 Tozer Road, Beverley, MA 01915-5599, USA.

Perkin–Elmer Cetus, 761 Main Avenue, Norwalk, CT 06859, USA.

Plasco Inc., 10 Jefferson Ave., Wolburn, MA, USA.

Promega, 2800 Woods Hollow Road, Madison, WI 53711-5399, USA. *Also:* Epsilon House, Enterprise Road, Chilworth Research Centre, Southampton SO1 7NS, UK.

Sarstedt Ltd., 68 Boston Road, Leicester LE4 1AW, UK.

Schleichel & Schuell, PO Box D-2254, Dassel, Germany

Sigma Chemical Co., Fancy Road, Poole, Dorset BH17 7TG, UK. *Also:* PO Box 14508, St Louis, MO 63178, USA.

Stratech Scientific Ltd., 50 Newington Green, London N16, UK.

Strategene Ltd., Cambridge Innovation Centre, Cambridge Science Park, Milton Road, Cambridge CB4 4GF, UK. *Also:* Stratagene Cloning Systems, 11099 North Torrey Pines Road, La Jolla, CA 92037, USA.

Techmate Ltd., 10 Bridgeturn Ave., Old Wolverton, Milton Keynes MK12 5QL, USA.

Techne Ltd., Duxford Road, Cambridge CB2 4PZ, UK.

US Biochemical Corporation, Box 22400, Cleveland, OH 44122, USA.

Contents of volume I

Index